SOILS AND
THE ENVIRONMENT

SOILS AND
THE ENVIRONMENT

A Guide to Soil Surveys
and their Applications

Gerald W. Olson

A DOWDEN & CULVER BOOK

CHAPMAN AND HALL
NEW YORK   LONDON

First published 1981 by
Chapman and Hall
733 Third Avenue, New York NY 10017

Published in Great Britain by
Chapman and Hall Ltd
11 New Fetter Lane, London EC4P 4EE

Printed in the United States of America

ISBN 0 412 23750 4 (cased edition)
ISBN 0 412 23760 1 (paperback edition)

Library of Congress Cataloging in Publication Data

Olson, Gerald W.
  Soils and the environment.

  Bibliography: p.
  Includes index.
  1. Soil-surveys. I. Title.
S592.14.037   641.4'7   81-10067
ISBN 0-412-23750-4   AACR2
ISBN 0-412-23760-1 (pbk.)

*This book is dedicated to my family — Mary,
Bradford, David, Eric — for their encouragement
and understanding.*

FOREWORD

As we enter the last decades of the twentieth century, many persistent and perplexing problems continue to afflict humankind. Thus it is appropriate to address, in a new group of books, two of the monumental issues that haunt people throughout the world. *Soils and the Environment* by Professor Gerald W. Olson is the first book in this new publishing program on Environment, Energy, and Society. The purpose of all these books will be to explore the many interrelated facets of these topics and to provide guidance for dealing with problems and offering ideas for their solutions. Environment and energy are twin problems that occupy what many believe to be opposite sides of a two-headed coin. They are often viewed as being antithetical and incompatible. The various books in this program will try to place in perspective the options that are available to those who design policy and plan and manage societal matters. Typical of books being developed currently are ones on coal resources, environmental geoscience, environmental pollution, land-use planning, nuclear energy, mineral resources, and water resources. However, because soils are at the very heart of civilization and provide the building block for human sustenance, it is fitting to inaugurate this series with Dr. Olson's timely analysis of soils. Unfortunately, these most vital resources seem to have low priority in many farming enterprises, urbanization projects, deforestation schemes, and mining and developmental terrain changes. Because of this almost universal lack of optimum management, understanding, and stewardship, these very fragile resources, the soils, are becoming likely candidates for the endangered species lists. The insights and strategies that Dr. Olson provides can go a long way toward helping to correct this ignorance.

We are indeed fortunate to have persuaded Professor Olson to write this book. His extraordinary wide experiences in analyzing soils throughout the world have given him unusual competence in the discipline of soil science. He has been working in research, teaching, and extension programs in soils investigations at Cornell University since 1962. His numerous publications attest to his expertise and ability to interpret soils under vastly differing climates and terrains. He has worked as a scientist and consultant for numerous political and governmental agencies, such as the United Nations, U.S. Agency for International Development, Kansas Geological Survey, and others. His many foreign assignments have included investigations in Australia, Brazil, El Salvador, Guatemala,

Honduras, India, Iran, Italy, the Philippines, Turkey, and Venezuela. Broadly based as it is on such worldwide knowledge, we are sure you will find *Soils and the Environment* an outstanding example of how to cure the present understanding and knowledge gap involving these most precious materials.

DONALD R. COATES

PREFACE

Soil surveys are being made and published in many areas, but too often they are not used to the fullest extent. This book provides a base for the expanded application of soil maps and reports. This publication gives an introduction to soil information through soil profile descriptions and soil mapping in landscapes; illustrates principles of grouping soils for practical purposes; and discusses some specific applications, including engineering interpretations, land classification, erosion control, yield correlations, archeological considerations, and aspects of using soil information in planning for the future. Additional and improved considerations for uses of soils in landscapes will help greatly in increasing the long-term productivity and performance of soils, and in protecting the environment from further abuse. Soils are a fragile part of the ecosystem, and must be managed carefully to achieve a harmonious environment—with human beings as an integral part of the efficient system. If soils are abused, the consequences can be disastrous. This book shows some examples of land abuse and the results, but concentrates principally on the scientific soil description, delineation, interpretation, and application of the information in a constructive format. Technology and examples are available for us to apply to achieve tremendous improvements in land use. The future can be bright indeed if we use our soils and other resources wisely.

Laypersons are the prime audience intended for this book, but the book will be equally useful to soil scientists and other technically trained people. The text of the book is constructed around the figures and tables, so that the basic message of the book is visual, based on the photographs, maps, diagrams, and graphs. The data in the tables provide detail for those who can benefit from specific examples of soil variability and significance of soil performance. Complex mathematical formulas and other highly technical discussions have been avoided, as have some of the more theoretical aspects of pedology. Selected references are listed to guide interested readers to further explanations about specific topics. A glossary is included for laypersons who may be unfamiliar with some of the terms.

This book can be of value to everyone involved or interested in uses of areas of land. In the ultimate sense, practically all human activities are concerned with or located on plots of different soils. Success or failure of foundations, farming operations, forestry, recreation, waste disposal, and nearly all other plot or area-dependent enterprises is greatly dependent upon the nature and properties of the soils. Contractors, for example, can use this book as the key to making interpretations about the uses of soil surveys for construction, excavations, and landscape design. Farmers using this book can improve their understanding of their soils and yields, and management considerations. Teachers and students will find in this book explanations about soil descriptions, classification, and groupings of soils for practical purposes. Environmentalists and hydrologists will obtain from the book an understanding of landscapes from which better decisions about ecological and pollution problems can be made in the future. Officials of the Soil Conservation Service and Cooperative Extension will discover that the book is useful as a reference for their own use and in teaching their clients and the general public about soils. Workers in international programs and natural scientists in others countries can use the book to generate interest in soil surveys, and to make better use of the surveys and reports that already exist. The book also relates to other disciplines, such as plant and animal ecology, land-use planning, physical geography, natural resources, earth resources, archaeology, agriculture, forest and range managment, environmental quality, resource economics, conservation, and many other fields. Specifically, this book will be used by agriculturalists, agronomists, assessors, botanists, conservationists, contractors, ecologists, economists, engineers, extension workers, foresters, geologists, groundwater experts, planners, politicians, public health officials, range managers, recreationists, soil scientists, students, teachers, wildlife specialists, and many others.

Many people have contributed over the past 25 years to the ideas and concepts expressed in this book. The author is especially grateful to reviewers Dr. Gerald A. Nielsen, Professor of Soils, Department of Plant and Soil Science, Montana State University, Bozeman, and Dr. Tej S. Gill, Senior Program Manager, Development Support Bureau, Office of Agriculture, Agency for International Development, Department of State, Washington, D.C. This book was written at the invitation of Dr. Donald R. Coates, Professor of Geology, Department of Geological Sciences and Environmental Studies, State University of New York at Binghamton, for his series, Environment, Energy, and Society, and the author greatly appreciates his encouragement and editorship. Sandra S. Seymour typed the manuscript, and Eileen W. Callinan drafted the illustrations.

G. W. OLSON

CONTENTS

CHAPTER 1

INTRODUCTION

Soils are basic to civilization and with water constitute society's most important resources. They provide food and fiber, support buildings and roads, help convert sunlight
into usable forms of energy and other resources. They are intimately connected with
geologic materials beneath, vegetative materials within and above, and groundwater percolating through the soil. Soils are an integral and vital part of our environment and may be
defined as discrete bodies produced by interactions of climate, vegetation, and surficial
geologic materials on the earth's surface. Soils vary greatly from place to place in landscapes, often within even short distances. The purpose of this book is to provide a key to
understanding and interpreting soil maps and reports that describe and delineate the
distribution of soils in landscapes. Better understanding of soil surveys will help to
improve the uses of them, and improving their uses will in turn result in real environmental enhancement for the benefit of everyone.

Soil survey reports appear in many forms in different countries, but the common
form of those recently published in the United States is illustrated in Figure 1. Detailed

FIGURE 1 *Examples of recent soil survey reports from different counties in
the United States.*

FIGURE 2 *Detailed Soil Map Sheet Number 23 of the area southeast of Ithaca, New York, reduced from the soil survey of Tompkins County (Neeley et al., 1965).*

soil maps are made in the field by digging in the soils and delineating their boundaries on aerial photographs. Commonly, soils are surveyed at a scale of 4 inches to 1 mile (1:15,840), and published at a scale of 1:20,000 — but the scales of the field survey and the soil map publication depend somewhat upon the complexity of the landscape and the intensity of land use in the survey area. General soil maps are also included to give users a bird's-eye view of soils in a survey area. The soil survey reports (Fig. 1) contain the soil maps and descriptions of the soils as well as interpretations about the uses of the soils for different purposes. Soils in the United States are mapped through the Cooperative Soil Survey — a joint effort of the U.S. Department of Agriculture, the Agricultural Experiment Station, and other agencies in each state.

Figure 2 is an example of a detailed soil map sheet. The area in the lower part of the valley of Sixmile Creek surrounding the Ithaca Reservoir has soils formed in calcareous clayey lacustrine (lake) sediments on steep slopes. Upland soils around Hungerford Hill and south of Ithaca College have mostly formed in compact glacial till, and some areas at higher elevations are shallow and moderately deep to bedrock. Symbols within each boundary identify the soil, slope, and erosion conditions.

In producing soil maps, soil scientists dig many holes in the landscape to make the delineations, and study soil color, texture, structure, porosity, consistence, pH, organic matter, kind of clay, and other profile characteristics (Fig. 3). They combine these observations with their knowledge of geomorphology, topography, vegetation, land use, and other landscape attributes to map different soil areas on aerial photographs. In a day of hard work, about 300 acres can be surveyed in the complex landscape at the detail of the soil map illustrated in Figure 2. In Figure 4 are shown some of the soil boundary conditions that may be visible in a freshly plowed field. The dark areas in the drainageway are poorly drained Madalin soils; somewhat poorly drained Rhinebeck soils are in the foreground; and some of the convex knolls in the background are occupied by moderately well drained Hudson soils. This area was mapped near the Cornell University Agricultural Experiment Station shown in the northern part of Soil Map Sheet Number 23 in Figure 2. Usually, soil boundaries are not as visually apparent as these, and soil scientists must dig a considerable number of holes to determine the internal characteristics of the soils in order to map them.

The main base of the soil mapping operation is the soil profile description. Figure 5 illustrates a soil pit from which a description of a soil profile can be made. Each soil has a unique combination of horizons with different properties that can be described and quantified. Soil profile descriptions are normally provided in the middle parts of soil map units in landscapes, and represent typical soil conditions of each map unit. The "modal" soil is thus defined and delineated within each landscape segment of a soil survey area. Each soil is defined on the basis of properties that make it unique for use and management of those landscape areas — so that the soil survey has a very practical goal for the mapping efforts. As a soil survey progresses, many soil profile descriptions are compiled. Examinations of auger corings and spade and shovel excavations are also used to supplement the soil profile descriptions. Ranges in soil characteristics are also described for each soil map unit, so that the variations from the "modal" soil within each landscape unit can be isolated and identified. If large spots within a map unit are significantly different from the defined soil, those spots must be identified and mapped as separate soils if the differences are great enough to affect use and management of the areas. Inclusions of different soils are also allowed in map units if the areas are too small to be mapped and if the soil differences do not appreciably affect use of the areas.

FIGURE 3 *Soil scientist examining a Langford soil developed in glacial till. The campus of Cornell University, built on soils developed mostly in glacial lacustrine sediments, is visible in the distance.*

FIGURE 4 *Soil boundaries visible in a freshly plowed field near the Cornell University Agricultural Experiment Station. These soils developed in calcareous clayey lacustrine sediments, and are mainly separated by the different drainage conditions.*

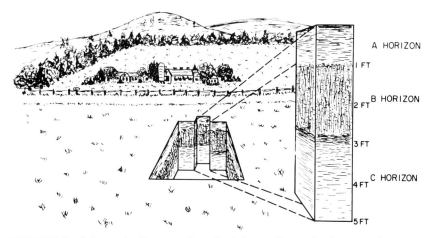

FIGURE 5 *Schematic diagram of a soil pit in a soil map landscape unit, from which the soil profile description area has been expanded to give a better view.*

FIGURE 6 *Soil scientists examining a roadcut with soil profile exposures during a tour of the American Society of Agronomy, near Los Angeles, California.*

Soil profiles and soil profile descriptions are of considerable interest to soil scientists, and they should also be of interest to all people who are involved in uses of the land. Figure 6 shows people examining lamellae (bands) of finer-textured materials in a roadcut near Los Angeles. The nature of these marine deposits affect soil behavior and influences stability of the soils during erosion and landslides common to the area. The success of both rural and urban developments is highly dependent upon planning that recognizes the proper use of soil resource information.

CHAPTER 2

SOIL PROFILE DESCRIPTIONS

Soils are different. The differences may range from very striking texture variations to more subtle color variations. Soil differences dictate that the soils must be managed differently and will behave differently when used for agriculture, forestry, sewage disposal, foundations, pavements, and other purposes.

The main basis for classification and understanding soils is the soil profile description made in the field, where soils occur. A soil profile is a vertical cut exposing the various parts of the soil. Laboratory investigations only supplement understanding of soil profile descriptions made in the field. Soil mapping is based on landscape occurrence of similar

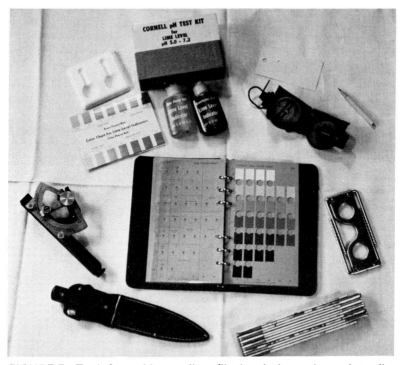

FIGURE 7 *Tools for making a soil profile description and mapping soils. These items are relatively inexpensive, and can be obtained by anyone interested in learning more about soils and their description and use.*

soil profiles and represents an actual area of soil (segment of the landscape). A soil profile description is normally made in a vertical cut, pit, or trench.

Soil profile descriptions are often interpreted and used by people without extensive experience in working with the soil survey. Thus, public health officials investigating septic tank seepage fields, planners of developing urban areas, extension service workers, construction and highway engineers, farmers and ranchers, foresters, conservationists and environmentalists, students beginning training in soils, teachers, and many others need to be able to understand the different soils they encounter in their work. The following information provides a convenient reference for the mechanics of making and interpreting a soil profile description.

Scattered sources of information for producing a soil profile description include the Soil Survey Manual (Soil Survey Staff, 1962), Soil Taxonomy (Soil Survey Staff, 1975), and other sources. A compilation of criteria for making and interpreting a soil profile description by Olson (1976) contains more detail than is given here. The procedure for constructing and interpreting a soil profile description consists simply of comparing properties of an individual soil profile with descriptive standards that have been established for these various properties for all soil profiles. Obviously, some people will have greater skill in developing and interpreting descriptions than others. Anyone with a willingness to learn can compare properties of soils with the clearly defined standards, and thus create and interpret a soil profile description.

Some of the tools for making a soil profile description and mapping soils are illustrated in Figure 7. The Munsell Soil Color Chart is a booklet of standard color chips by which the colors of small soil samples can be identified. The pH test kit has indicators for

TABLE 1 *Form for describing environment around the site (cut, pit, or trench) for a soil profile description (Soil Survey Staff, 1962)*

SOIL TYPE		FILE NO.
AREA	DATE	STOP NO.
CLASSIFICATION		
LOCATION		
VEGETATION (OR CROP)	CLIMATE	
PARENT MATERIAL		
LANDFORM		
RELIEF	DRAINAGE	SALT OR ALKALI
ELEVATION	GROUND WATER	STONINESS & ROCKINESS
SLOPE	MOISTURE	
ASPECT	ROOT DISTRIBUTION	
EROSION		
PERMEABILITY		
ADDITIONAL NOTES, PHOTOS, ETC.		

colorimetric measurement of soil pH in the field on small samples within a few minutes. A knife facilitates digging in the exposed soil profile to collect samples and examine structure and consistence. Plastic bags and tags enable collection of samples; a pencil is used to take notes. The Abney level is used to measure slopes, the compass to take azimuth directional readings, and the pocket stereoscope enables determination of topographic differences when aerial photographs are viewed in stereo pairs. Soil horizon thickness is measured with a folding rule or tape measure. Pocket microscopes and thermometers are used to examine small features in soils and to determine some of the climatic characteristics of soils. These items are relatively inexpensive, and can be useful to everyone interested in describing soils and interpreting soil survey information.

The Soil Conservation Service has prepared a form for making soil profile descriptions (Soil Survey Staff, 1962). This form consists of two parts: (1) a description of the environment in which the soil occurs, and (2) a detailed soil profile description of a vertical section of the representative soil from the segment of the landscape being studied or mapped, made from a vertical cut, pit, or trench. Table 1 illustrates the form for description of the environment. Table 2 gives the form for description of the various horizons into which the soil profile is subdivided. Conscientious description of all the items in Tables 1 and 2 explains the landscape and the soil quite comprehensively, in adequate detail for most interpretations. From this basic record of soil and environmental data much useful information can be derived to indicate how the soil can best be used. The described soil factors mapped on aerial photographs show:

1. Landform, relief, drainage
2. Parent material of soils, geology
3. The soil profile
    a. Color
    b. Texture
    c. Structure
    d. Porosity
    e. Consistence
    f. Acidity, alkalinity, lime status
    g. Concretions, other special formations
    h. Organic matter, roots
    i. Chemical and mineralogical composition
    j. Other characteristics
4. Stoniness and rockiness
5. Erosion
6. Vegetation
7. Land use
8. Other significant features

The assembly of data recorded for a soil profile in Tables 1 and 2 and their statistical or computer analyses may take many forms. Table 3 illustrates the typical style of organization of the information. This is the style in which the soil profile description appears in the technical description of a published soil survey report (Neeley et al., 1965). The Rhinebeck soil described in Table 3 is mapped in the RkA and RkB areas shown in Figure 2, and the Rhinebeck landscape and soil surface is shown in Figure 4. Because of the high clay content, the soil is slowly permeable and has many problems with wetness and frost heaving.

TABLE 2 *Form for describing horizons of a soil profile (Soil Survey Staff, 1962)*

Horizon	Depth	Color		Texture	Structure	Consistence			Reaction	Boundary		
		Dry	Moist			Dry	Moist	Wet				

TABLE 3 *Soil profile descripton of Rhinebeck silt loam from roadcut along Pine Tree Road ½ mile east of Ithaca, New York*

Horizon	Depth (in.)	Description
Ap	0–8	Very dark grayish brown (10YR 3/2) when moist; silt loam; moderate medium granular; friable; nonplastic; many fine and few medium roots; pH 5.6; clear smooth boundary; 7 to 9 in. thick
A2	8–13	Grayish brown (10YR 5/2) when moist; silt loam; common medium distinct (10YR 5/4) mottles; weak thin platy; friable; nonplastic; few fine and medium roots; pH 5.2; gradual wavy boundary; 3 to 7 in. thick
B21	13–18	Brown (10YR 4/3) when moist; silty clay loam; few faint medium mottles; moderate medium subangular blocky; prominent clay coats; ped interiors brown (10YR 5/3) with common medium distinct (10YR 5/6) mottles; firm; plastic; sticky; few medium roots on ped faces; pH 5.5; gradual smooth boundary; 4 to 6 in. thick
B22	18–28	Brown (10YR 4/3) when moist; silty clay; few fine faint mottles; weak coarse angular blocks crushing to strong medium and fine subangular blocks; clay coats prominent; interiors brown (10YR 5/3) with many medium distinct (10YR 5/6) mottles; firm; very plastic; sticky; few medium roots on coarse block faces; pH 6.2; clear smooth boundary; 8 to 14 in. thick
C	28–36+	Pale brown (10YR 6/3) moist; silty clay loam; strong medium and thick platy; firm; plastic; slightly sticky; many lime nodules $1/16$ to $1/4$ in. in diameter; no roots; pH 8.0; strong effervescence with HCl; lower boundary not seen

TABLE 4 *Notations for use with the form in Table 1 for describing the environment around a soil profile description site*

Soil type — Name, as Rhinebeck silt loam, plus soil map unit symbol, if any

Classification — Name in lowest category known in Soil Taxonomy, as Aeric Ochroqualf, fine, illitic, mesic (Soil Survey Staff, 1975)

Climate — Such as: humid temperate, warm semiarid, or other climatic information or classification of area

Parent material — Such as: residuum from basalt, mixed silty alluvium, calcareous clay loam till

Landform — Such as: high terrace, till plain, alluvial fan, mountain foot slope. Add name of geologic formations, where known

Relief — Give letter designation of soil map unit or name of soil slope class (level, sloping, steep) and indicate concave or convex, simple or complex slopes

Slope — Give gradient of soil slope from Abney level measurement

Aspect — Give direction of slope and wind and sun effects on soil

Erosion — Use appropriate class name and number; as slight, moderate, or severe erosion

Permeability — Give rate: as rapid, moderate, or slow

Drainage — Use appropriate class name for soil drainage: as excessively drained, well drained, moderately well drained, somewhat poorly drained, poorly drained, or very poorly drained

Groundwater — Give depth to groundwater or indicate approximate depth

Moisture — Indicate present soil moisture: as wet, moist, moderately dry, or dry

Root distribution — Indicate depth of penetration, size and kind of roots, and abundance as few, common, or many

Salt or alkali — Indicate concentration: as slight, moderate, or strong

Stoniness or rockiness — Use appropriate name: as stony, very stony, extremely stony, or rocky, very rocky, extremely rocky

Additional notes, photos, etc. — Include additional detail on listed items or include additional items, such as relative content of organic matter, evidence of worms, insects, or rodents, special mottling, stone lines, and other features with photos and sketches to indicate soil conditions

TABLE 5 *Notations for use with the form in Table 2 for describing horizons of a soil profile*

Horizon:	O, A, B, C, R
Depth:	From top of soil surface
Color:	Matrix and mottles dry and moist
Mottling:	Spots of different colors

Abundance	Size	Contrast
Few—less than 2%	Fine—less than 5 mm	Faint
Common—2–20%	Medium—5–15 mm	Distinct
Many—more than 20%	Coarse—more than 15 mm	Prominent

Textures: Size of mineral particles
Sands: very coarse sand, coarse sand, sand, fine sand, very fine sand
Loamy sands: Loamy coarse sand, loamy sand, loamy fine sand
Sandy loams: Sandy loam, fine sandy loam, very fine sandy loam

Silt	Silty clay loam	Silty clay
Silt loam	Clay loam	Clay
Loam	Sandy clay loam	Sandy clay

Adjectives: Gravelly, cobbly, channery, flaggy, stony may apply to any of these

Structure: Aggregation of soil particles
Class: Size in mm

	Prisms	Blocks	Plates and Granules
Very fine (very thin)	0–10	0–5	0–1
Fine (thin)	10–20	5–10	1–2
Medium	20–50	10–20	2–5
Coarse (thick)	50–100	20–50	5–10
Very Coarse (very thick)	100+	50+	10+

Grade: Distinctness
    Structureless, very weak, weak, moderate, strong, very strong
Type: Form
    Prismatic, columnar
    Blocky, angular blocky, subangular blocky
    Granular, crumb
    Platy

Consistence: Cohesion and adhesion and resistance to deformation
    Dry: Loose, soft, slightly hard, hard, very hard, extremely hard
    Moist: Loose, very friable, friable, firm, very firm, extremely firm
    Wet: Nonsticky, slightly sticky, sticky, very sticky
        Nonplastic, slightly plastic, plastic, very plastic

Reaction:	Use pH figures
Boundary:	Transition from one horizon to another

Distinctness		Topography
Abrupt	0–1 in.	Smooth—nearly a plane
Clear	1–2½ in.	Wavy or undulating—pockets wider than deep
Gradual	2½–5 in.	Irregular—pockets deeper than wide
Diffuse	5+ in.	Broken—parts unconnected

A soil profile description—properly made, understood, and interpreted—supplies a wealth of information about the internal and external characteristics of the segment of the landscape with which it deals. By studying the soil profile description and accompanying data, one can tell what the landscape looks like and what its subsurface composition is. Interpretations can be made for all uses to which the soils must be put. Maps made by classifying similar soil profiles of a landscape show which soils are located in each area. Profile descriptions representing soils from different places can be compared and classified—enabling each soil to be better understood in the natural order of things.

Tables 4 and 5 list some types of notations for use with the forms in Tables 1 and 2. The procedures for preparing soil profile descriptions have been well established through many years of experience in producing soil surveys, but the current procedures will be modified as our knowledge of soils increases. Those who describe and use soils should not be entirely bound by convention, but should use their initiative freely to describe special and significant soil features in terms understandable to other scientists and laymen. The use of special tools (such as photography) should be fully employed for description and study of soils, and for communication of soils information to others.

SOIL HORIZONS

The composition and arrangement of soil horizons in a profile are the major determinants in the classification, mapping, and use of land areas. The description of a soil profile consists mainly of individual examination of its separate horizons. Table 6 lists and briefly describes symbols used to designate soil horizons. Figure 8 illustrates soil horizons in a pit prepared for soil profile description and sampling; the horizons in the pit have been marked to show their boundaries. A soil horizon is a layer of soil approximately parallel to the soil surface with characteristics influenced by genetic processes. Each horizon is separated from adjacent ones on the basis of differences in properties. The soil profile, as exposed in a road cut or pit, includes the collection of all the genetic horizons, the natural organic layers on the surface, and the geologic materials that influence the genesis and behavior of the soil. The profile that represents soil map units must be characteristic of large areas of the same kind of soil.

In describing a soil profile, one usually locates the boundaries between horizons, measures their depth, and studies the entire profile before describing and naming individual horizons. The blanks in the form in Table 1 are filled out at this time.

Designation of soil horizons, listed in Table 6, according to the descriptive format of Table 2, enable interpretations for the genesis of a soil. An understanding of the genesis contributes important knowledge for interpretations about the use of soil areas. Thus, recent alluvial sediments, volcanic ash deposits, and windblown (loess) materials have soils that are relatively young with poorly developed B horizons. Older uneroded soils are thicker in more humid tropical regions than in temperate areas because they have been more deeply weathered. The experienced soil scientist making a soil profile description has the best knowledge of the soil in its own environment at that moment of description, and the soil profile description, together with the soil horizon designations, represents the best field assessment of the soil properties. All subsequent laboratory analyses constitute only accessory data, and all data interpretation to be accurate must consider the field context of the soils (the soil profile descriptions). The soil profile description constitutes the most important and basic data that can be gathered about a soil.

TABLE 6 *Designations of master horizons and subordinate symbols for horizons of soil profiles (adapted from Soil Survey Staff, 1962, with 1962 Supplement)*

Master horizons

01	Organic undecomposed horizon
02	Organic decomposed horizon
A1	Organic accumulation in mineral soil horizon
A2	Leached bleached horizon (eluviated)
A3	Transition horizon to B
AB	Transition horizon between A and B—more like A in upper part
A and B	A2 with less than 50% of horizon occupied by spots of B
AC	Transition horizon, not dominated by either A or C
B and A	B with less than 50% of horizon occupied by spots of A2
B	Horizon with accumulation of clay, iron, cations, humus; residual concentration of clay; coatings; or alterations of original material forming clay and structure
B1	Transition horizon more like B than A
B2	Maximum expression of B horizon
B3	Transitional horizon to C or R
C	Altered material from which A and B horizons are presumed to be formed
R	Consolidated bedrock

Subordinate symbols

b	Buried horizon
ca	Calcium in horizon
cs	Gypsum in horizon
cn	Concretions in horizon
f	Frozen horizon
g	Gleyed horizon
h	Humus in horizon
ir	Iron accumulation in horizon
m	Cemented horizon
p	Plowed horizon
sa	Salt accumulation in horizon
si	Silica cemented horizon
t	Clay accumulation in horizon
x	Fragipan horizon
II, III, IV	Lithologic discontinuities
A'2, B'2	Second sequence in bisequal soil

Horizon designations (Table 6) are symbols indicating the judgment of the experienced soil scientist, who describes the type and degree of soil departure from the original geologic material. Each symbol indicates the best judgment of the describer, who is the most qualified to describe the soil at that moment in the field. The symbol and descriptive nomenclature implies that readers and interpreters of that soil profile description must mentally reconstruct the character of the original geologic material and the soil that formed in it.

Master horizons of soil profiles are designated as O, A, B, C, and R (Table 6). O horizons have formed above the mineral surface of soils, are dominated by fresh or partly decomposed organic material, and contain more than 30 percent organic matter if the mineral fraction is more than 50 percent clay, or more than 20 percent organic matter if

FIGURE 8   *Soil profile horizons marked off in a pit for description and sampling. This soil in New York State is being studied as part of the Cooperative Soil Survey program for soil characterization.*

the mineral fraction has no clay (or proportional organic matter content with intermediate clay content). The O horizons are generally present at the soil surface under native vegetation, but may have been buried by sedimentation (alluvial, loess, or ash fall). When plowed, O horizons are generally destroyed (incorporated into the Ap). O1 horizons have the original form of most vegetative matter visible to the naked eye; O2 horizons are relatively decomposed, and the plant materials cannot be recognized with the naked eye. Organic soils in swamps and bogs have a separate nomenclature for their description, based on the characteristics of the wet organic layers.

A horizons are mineral horizons with organic matter accumulations beneath the soil surface. These horizons have also generally lost clay, iron, or aluminum with resultant concentration of quartz or other resistant minerals of sand or silt size, and are transitional to underlying B or C horizons. A1 horizons are dominated by an accumulation of dark humified organic matter associated with the mineral fraction. A2 mineral horizons are dominated by loss of clay, iron, or aluminum (eluviation), with resultant concentrations of quartz or other resistant minerals in sand and silt sizes. Commonly, A2 horizons are lighter in color and lower in organic matter than overlying A1 horizons and underlying B layers. An A3 horizon is transitional between A and B, but dominated by properties characteristic of the overlying A1 or A2 (but having some subordinate properties of an underlying B). AB horizons are also transitional, but are usually too thin to be conveniently separated into A3 and B1. "A and B" horizons include A2 material with less than 50 percent B; these appear to be mostly in soils where the A2 is encroaching on an underlying B. AC horizons are thin, transitional between A and C, but not dominated by properties characteristic of either A or C.

B horizons (Table 6) have an illuvial concentration (moved from above eluvial horizons by leaching waters) of silicate clay, iron, aluminum, or humus, a residual concentration of sesquioxides or silicate clays that has formed by means other than solution and removal of carbonates or more soluble salts, coatings of sesquioxides giving darker or redder colors, or alteration of parent materials to form significant structure in subsoils. B1 horizons are transitional from A1 or A2, but more like the underlying B2. "B and A" horizons include A2 material with more than 50 percent B; these horizons commonly have tongues of A2 extending into the upper part of the B horizon. B2 horizons have the maximum expression of accumulation of clay, iron, humus, color, or structure moved in an illuvial fashion and concentrated in the subsoil. The B3 is a transitional horizon between B and C or R, but dominated by properties of the overlying B. The A and B horizons collectively are commonly called the solum of the soil.

C horizons have been relatively less affected by soil genetic factors. However, they are subject to weathering outside the zones of major biological activity, reversible cementation, development of brittleness, high bulk density, gleying, accumulation or cementation of calcium or magnesium carbonate or more soluble salts, or cementation by alkali-soluble siliceous material or by iron and silica. C horizons have been some affected by soil-forming processes, but do not show the effects of biological and pedological activity as much as do the A and B horizons.

R horizons are the underlying consolidated bedrock, such as granite, sandstone, or limestone. The symbol R used alone presumes that the material is the parent rock of the soil. An R horizon is generally composed of hard rock (lithic contact with overlying soil) but may also be soft rock (paralithic contact with overlying soil) such as marl or soft shale.

Subordinate symbols are used as suffixes after master horizon designations to indicate dominant features of different kinds of master horizons, as indicated in Table 6. Generally, only one subordinate symbol is used, and subordinate symbols are specific to certain master horizons.

Horizons may be subdivided arbitrarily for sampling and description if they are thick; B2 horizons may be designated, for example, as B21, B22, and B23. Certain conventions are used for combinations of the symbols; thus, typical correct combinations of letters and numbers are B21t, C1g, C2g, Ap1, and Ap2. The most important characters in the descriptive portions of the symbol are generally placed first.

Lithologic discontinuities of soil material from different sources are indicated by Roman numerals added to the horizon designation. Two examples of horizon sequences using this convention are:

A1-A2-B1-B21-IIB22-IIB3-IIC1-IIIC2

A1-A2-B1-B2-IIA2-IIBx-IIC1x-IIIC2x-IIIC3-IVR

Lithologic discontinuities are common in landscapes where erosion is severe. Often, the discontinuities are accompanied by stone lines from previous erosion cycles. In some places soils have developed several sequences of A and B horizons; these may be superimposed one over another to constitute bisequal or trisequal soil profiles. Bisequal soils are likely the result of climate changes acting on uniform geologic materials, or are the remnants of former soil profiles (paleosols) which have been buried under younger soils. Some places in the Midwestern United States have three or four separate paleosol formations.

Profiles of soils having well-developed microrelief sometimes cannot be satisfactorily described by study of a single pit. Instead, if a trench is dug in these places, the soil variations are more fully exposed and can then be easily characterized. Often horizontal patterns can be exposed in soils by removing horizons from the top down from an area of 1 square meter or more. Depth and thickness of soil horizons must be carefully observed and recorded in Table 2. Range in thickness is conventionally given for each horizon in each description (Table 3) made from a pit or other soil exposure, but when a soil survey report is published, the range in thickness is given for horizons of all the soils in a survey area. Some horizons characteristically tongue into the horizons below them, so that these boundary conditions need to be specially described from extensive excavations.

COLOR

Soil colors are important to observe and record because they provide information about the properties and behavior of each soil. Reddish colors often indicate soils highly weathered with large oxidized iron content; grayish colors may be diagnostic of soils with permanently high and stagnant water tables. Mottled soils with spots of different colors in humid regions may have fluctuating water tables; the mottles (colors) in the soils may indicate the seasonal wetness conditions even when the soils are examined during the drier seasons. Soil colors are properties that can be easily observed and measured, and the colors enable inferences to be made about soil mineralogy, stage of weathering, organic matter content, seasonal water fluctuations, and many aspects of soil performance and use.

Each soil profile may consist of several or many horizons differing in color. The complete color description should be recorded for each horizon under both dry and moist conditions in the field when a soil profile is examined, as outlined in the form in Table 2. Soil colors are most conveniently measured by comparing small soil samples with the chips on a color chart (Fig. 7). The color booklet usually used to describe soil color is called a Munsell Soil Color Chart. It has several hundred different colored chips systematically arranged on pages by hue, value, and chroma—the three simple variables that combine to give all colors. Hue is the dominant spectral (rainbow) color; it is related to the dominant wavelength of the light. Value refers to the relative lightness of color

TABLE 7 *List of Munsell Soil Color Chart pages with ranges of soil color names used to describe soil horizons (adapted from Soil Survey Staff, 1962, 1975; Munsell Color Company, 1954; and Olson, 1976)*

5R	Gray to black; pale red to very dusky red; light red to dark red
7.5R	Gray to black; pale red to very dusky red; light red to dark red
10Y	Reddish gray to reddish black; pale red to very dusky red; light red to dark red
2.5YR	Gray to black; pale red to very dusky red; light reddish brown to dark reddish brown; light red to dark red
5YR	White to black; pinkish white to dark reddish brown; reddish yellow to yellowish red
7.5YR	White to black; pinkish white to very dark brown; reddish yellow to strong brown
10YR	White to black; very pale brown to very dark brown; yellow to dark yellowish brown
2.5Y	White to black; pale yellow to very dark grayish brown; yellow to olive brown
5Y	White to black; pale yellow to dark olive gray; yellow to dark olive
Gley	Light gray to dark gray; light greenish gray to dark greenish gray; light bluish gray to dark bluish gray; pale green to grayish green

and is a function (approximately the square root) of the total amount of light. Chroma (saturation) is the relative purity or strength of the spectral color and increases with decreasing grayness.

Table 7 provides a list of the page designations and color ranges in the Munsell Soil Color Chart. Figure 9 illustrates one page from the booklet, including the soil color names in their respective positions on the 5YR hue page. The colors on a page are of a constant hue, and the hues on the different pages are designated by such symbols as 5R, 7.5R, 10R, 2.5YR, 5YR, 7.5YR, 10YR, 2.5Y, and 5Y. Vertically, the colors on a page become successively lighter by visually equal steps as their value increases. Horizontally on a page, the colors increase in chroma to the right and become grayer to the left. Pages in the Munsell Soil Color Chart range from red (R) to yellowish red (YR) to yellow (Y) from the front to the back of the book (Munsell Color Company, 1954).

By convention, Munsell color names are recorded systematically and symbolically in a notation of hue, value, and chroma, such as 10YR 5/4 for yellowish brown. The nomenclature for soil color thus includes both the Munsell notation and the color names.

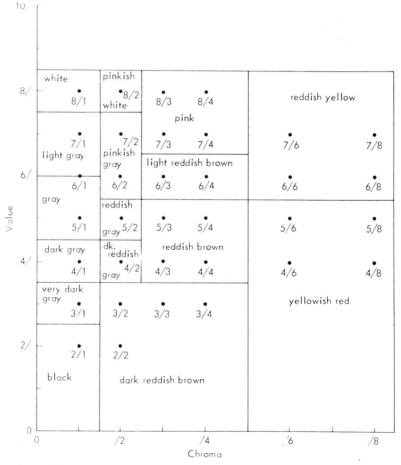

FIGURE 9 *Soil color names for combinations of value and chroma and 5YR hue (adapted from Soil Survey Staff, 1962, 1975, and Munsell Color Company, 1954).*

In the Munsell Soil Color Chart system, soil color components have three scales: (1) radial in hue, from one page to the next; (2) vertical in value; and (3) horizontal in chroma. The Munsell notation is particularly useful for international soil correlations because no translation of color names is needed from the symbolic recording of the Munsell notation. With a standard system, differences in partial color blindness or skewness and personal perspective can be eliminated or reduced when different soil scientists describe soils.

Color determinations in soils of a uniform matrix are relatively easy, but recording observations becomes more complex in multicolored horizons. Many soils have coatings or stains on faces of aggregates or within them. Often, it is useful to record colors of crushed and rubbed aggregates as well as colors within broken soil structural units. Mottling in soils is described by noting both the color of matrix and of mottles, together with size and proportion of mottles. Description of the pattern of mottles requires three sets of notations for (1) contrast, (2) abundance, and (3) size. The notations are summarized in Table 8.

Contrast of different mottles is faint when they are indistinct and recognizable only with close examination; mottles with closely related hues and chromas (compared to the matrix color) are classified as indistinct or faint. Although not striking, distinct mottles are readily seen, value and chroma are several units apart, and hue is one or two units different from the matrix color. Prominent mottles are very obvious and mottling is one of the outstanding features of the horizons. Hue, chroma, and value must be several units apart between the matrix and the mottle colors for prominent mottling.

Abundance of mottles (Table 8) is expressed as few, common, or many, depending upon the relative amount of mottled surface in the area exposed for description. Figure 10 is a chart that assists in estimating percentage proportions of mottles. Mottles have many different shapes in soils, and these should be described in special notations if they are significant. Close-up photographs and microphotographs can be useful to describe these special features of color combinations in soils.

Size classes of mottles are fine, medium, and coarse of <5, 5-15, or >15 mm diameter size ranges. Sometimes the boundaries of mottles need to be described as sharp (knife-edge), clear (<2 mm), or diffuse (>2 mm). Many mottles are roughly circular, but some are elongated or merge into streaks and tongues.

TABLE 8 *List of terms used to describe soil mottling (adapted from Soil Survey Staff, 1962)*

Contrast	
Faint	Indistinct mottles. Closely related hues and chromas of matrix and mottles
Distinct	Mottles easily seen. Mottles vary from matrix as much as 1 or 2 hues or several units in chroma or value
Prominent	Conspicuous mottles. Hue, value, and chroma may be several units apart
Abundance	
Few	Less than 2% of exposed surface
Common	2–20% of the exposed surface
Many	More than 20% of the exposed surface
Size	
Fine	Less than 5 mm in diameter
Medium	5–15 mm in diameter
Coarse	Larger than 15 mm in diameter

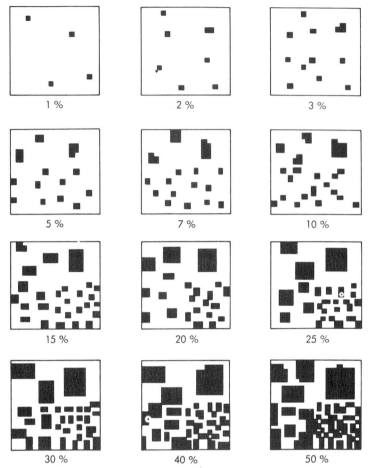

FIGURE 10 *Chart for estimating proportions of mottles (each fourth of any one square has the same amount of black; adapted from Folk, 1951, and Olson, 1976).*

Soil color changes with moisture content, more markedly in some soils than in others. Thus, both dry and moist colors should be described, and wet colors also in some cases. Moist soil colors are usually darker by ½ to 3 steps in value and differ by ½ to 2 steps in chroma from dry soils; seldom are they different in hue. Mottle color detail should not be excessively recorded, and good judgment is required to make accurate soil profile descriptions illustrating relative importance of all details of the color combinations.

Readings of soil colors should be made in standardized conditions of lighting as much as possible. Direct comparisons of soil samples with Munsell color chips usually give best results when the colors are exposed in open shade. Extremely bright sunlight or low light in dusk conditions in the field give less precise data. Munsell color chips fade over time, so they should be protected from bright sunlight, kept as clean as possible, and replaced periodically after prolonged use. With proper care and techniques, the Munsell color

TABLE 9 Simplified list of terms describing soil texture, coarse fragments, stones, and rocks (adapted from Soil Survey Staff, 1962, 1975, and Olson, 1976)

Separates	Diameter (mm)
Very coarse sand	2–1
Coarse sand	1–0.5
Medium sand	0.5–0.25
Fine sand	0.25–0.1
Very fine sand	0.1–0.05
Silt	0.05–0.002
Clay	Less than 0.002

General soil textures	Textural class names
Sandy soils—	
Coarse textured	Sand (Coarse, Medium, Fine, Very fine), Loamy sand (Loamy coarse sand, Loamy sand, Loamy very fine sand)
Loamy soils—	
Moderately coarse textured	Sandy loam (Coarse sandy loam, Sandy loam, Fine sandy loam, Very fine sandy loam)
Medium textured	Very fine sandy loam, Loam, Silt loam, Silt
Moderately fine textured	Clay loam, Sandy clay loam, Silty clay loam
Clayey soils—	
Fine textured	Sandy clay, Silty clay, Clay

Field determinations of texture

Sand—Loose and single grained.

Sandy loam—Moist cast bears careful handling. Many sand grains visible. Squeezed when dry it forms a weak cast that crumbles at a light touch.

Loam—Has a relatively even mixture of sands, silt, and clay. Some sand grains visible. Forms a rough broken ribbon between thumb and finger. Wet cast can be deformed slightly without crumbling. Dry cast bears careful handling.

Clay loam—Few sand grains visible. Forms a thin smooth slick ribbon that will bear its own weight. Wet cast can be molded into different shapes but tends to break as moisture is worked out.

Silt loam—Moist cast bears handling. Forms a rough broken ribbon that will not bear its own weight. Dry casts may be handled freely. Dry material feels like flour.

Silty clay loam—Moist cast bears handling. Forms a thin smooth ribbon that just bears its own weight. Wet cast can be kneaded into different shapes but with a tendency to crack as moisture is worked out.

Silty clay or Clay—Thin ribbon easily bears its own weight. Wet cast can be molded into different shapes without breaking.

Coarse fragment shape	To 3	Diameter (in.) 3–10	More than 10
Round	Gravelly	Cobbly	Stony or bouldery
Irregular	Angular gravelly	Angular cobbly	Stony
Thin flat	Channery	Flaggy	Stony

Stoniness Classes

0—No stones or too few to interfere with tillage

1—Stones interfere with tillage but intertilled crops can be grown

2—Stones make intertilled crops impracticable, but soil can be worked for hay or improved pasture

3—Stones make use of heavy machinery impracticable

4—Stones make use of all machinery impracticable

5—Rubble

Rockiness Classes

0—No bedrock exposures or too few to interfere with tillage

1—Bedrock exposures interfere with tillage but intertilled crops can be grown

2—Bedrock exposures make intertilled crops impracticable, but soil can be worked for hay or improved pasture

3—Bedrock exposures make use of heavy machinery impracticable

4—Bedrock exposures make all use of machinery impracticable

5—Rock outcrop

system enables excellent descriptions to be made of soil color combinations in forms that are readily understandable by other scientists and laypersons.

TEXTURE

Texture is the size and distribution proportion of mineral particles comprising soil horizons (Soil Survey Staff, 1962, 1975). Texture is generally the most permanent and probably the most important of all soil characteristics. Table 9 is a simplified list of terms describing soil texture, coarse fragments, stoniness, and rockiness.

Soil separates are the individual size groups of mineral particles below 2 mm in diameter. These size ranges are the most important for most chemical, physical, and mineralogical reactions in soils, and influence root growth for plants and the engineering behavior for structures. The finer particles are relatively more important reactively than the larger particles, which are relatively more inert. About 4 lb of dry clay particles with a diameter of 0.001 mm have a total surface area of about an acre (Soil Survey Staff, 1962). Larger particles and stones and rock fragments influence tillage and behavior of soils under roadbeds and foundations.

The size limits of soil separates are listed in Table 9. Percentage proportions by weight are determined and named according to the textural triangle shown in Figure 11.

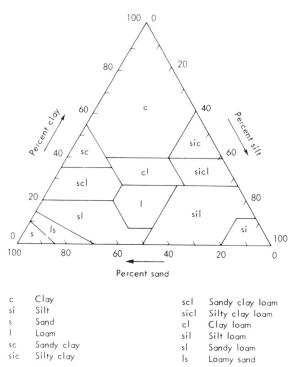

c	Clay	scl	Sandy clay loam
si	Silt	sicl	Silty clay loam
s	Sand	cl	Clay loam
l	Loam	sil	Silt loam
sc	Sandy clay	sl	Sandy loam
sic	Silty clay	ls	Loamy sand

FIGURE 11 *Textural triangle showing the percentage of clay (less than 0.002 mm), silt (0.002-0.05 mm), and sand (0.50-2.0 mm) in the basic soil textural classes (adapted from Soil Survey Staff, 1962, 1975).*

Sand particles are visible with the naked eye (Fig. 12) and have a gritty feel to the fingers. Silt particles (Fig. 13) cannot be seen without a microscope, but they have a distinctive nonsticky "smooth" feel to the fingers, like talcum powder, when dry or wet. Clay particles are even smaller than silt particles, are sticky when wet, and form compact lumps when dry.

FIGURE 12 *Sample of sandy soil composed mostly of sand particles, each of which can be seen with the naked eye. This soil sample was collected from an alluvial sandbar deposit near a stream.*

FIGURE 13 *Sample of silty soil composed mostly of silt particles too small to be seen with the naked eye. This soil sample was collected from a landslide area of unstable soils prone to sliding when they are wet. The sample has been partially crushed to destroy most of the structure for particle-size analyses.*

Field Definitions

Most soil horizons have different proportions of sand, silt, and clay — so the relative proportions must be determined in order to describe the texture. Field definitions have been devised (Shaw, 1928; Soil Survey Staff, 1962, 1975) to relate the basic soil textural classes as felt with the fingers in the field to the proportions graphed on the textural triangle (Fig. 11).

Sand is loose and single-grained. The individual grains can readily be seen or felt. Squeezed in the hand when dry it will fall apart when the pressure is released. Squeezed when moist, it will form a cast, but will crumble when touched.

Sandy loam is a soil containing much sand but which has enough silt and clay to make it somewhat coherent (Shaw, 1928; Soil Survey Staff, 1962, 1975). The individual sand grains can be readily seen and felt. Squeezed when dry, it will form a cast that will readily fall apart, but if squeezed when moist, a cast can be formed that will bear careful handling without breaking.

Loam is a soil having an apparent relatively even mixture of different grades of sand, silt, and clay. It is easy to manipulate with a somewhat gritty feel, yet fairly smooth and slightly plastic. Squeezed when dry, it will form a cast that will bear careful handling, but the cast formed by squeezing moist soil can be handled quite freely without breaking.

Silt loam (Shaw, 1928; Soil Survey Staff, 1962, 1975) is a soil having a moderate amount of fine grades of sand and only a small amount of clay, with over half silt-size particles. When dry it may appear cloddy, but the lumps can be readily broken, and when pulverized it feels soft, smooth, and floury. When wet, the soil readily flows together and puddles. Either dry or moist it will form casts that can be freely handled without breaking, but when moistened and squeezed between thumb and finger it will not ribbon but will give a broken appearance.

Clay loam is a fine-textured soil that usually breaks into clods or lumps that are hard when dry. When the moist soil is pinched between the thumb and finger it will form a thin ribbon that will break readily, barely sustaining its own weight. The moist soil is plastic and will form a cast that will bear much handling. When kneaded in the hand it does not crumble readily but tends to work into a heavy compact mass.

Clay (Shaw, 1928; Soil Survey Staff, 1962, 1975) is an ultra-fine-textured soil that usually forms very hard lumps or clods when dry and is quite plastic and usually sticky when wet. When the moist soil is pinched out between the thumb and fingers it will form a long flexible ribbon. Exceptions are found in some fine clays of tropical areas very high in colloids with good aggregation which are friable and lack plasticity in all conditions of moisture.

From the definitions (Shaw, 1928; Soil Survey Staff, 1962, 1975), other soil textures can be interpolated from the textural triangle (Fig. 11)—by relating the "feel" of the soil in the field to the textural triangle. Loamy sand (Soil Survey Staff, 1975) has about 85–90 percent sand, and the percentage of silt plus 1½ times the percentage of clay is not less than 15. Sandy clay loam has 20–35 percent clay, less than 28 percent silt, and 45 percent or more sand. Silty clay loam has 27–40 percent clay and less than 20 percent sand. Sandy clay has 35 percent or more clay and 45 percent or more sand. Silty clay has 40 percent or more clay and 40 percent or more silt (Soil Survey Staff, 1975).

A Practical Exercise

The value of the textural triangle (Fig. 11) is its educational merit in teaching people how to determine soil texture. As a class exercise, six moist soil samples of widely contrasing textures were brought into the classroom in plastic bags. Students were asked to determine the percent sand, silt, and clay of each sample with their fingers—by relating the descriptions (Shaw, 1928; Soil Survey Staff, 1962, 1975) to the textural triangle (Fig. 11). The class determinations are given in Table 10 for 17 students for Niagara, Canaseraga, Red Hook, Wayland, Arkport, and Tioga soil samples; the students included both graduate and undergraduate registrants, but few had any previous experience in determining field textures of soils.

From the data in Table 10, it is apparent that variability of percentage estimates are great, but that most people can do a relatively good job in determining soil texture even without extensive experience. Student 15 showed the best performance, with only 146 percentage units deviation from the laboratory data; Student 10, on the other hand, had 442 percentage units missed from the lab determinations. In general, the class as a whole did better in determining percentage sand content on the soils with relatively low sand (Niagara, Canaseraga, Red Hook); sand content was more difficult for the class to estimate in soils with more than 50 percent sand (Wayland, Arkport, Tioga). Silt content

TABLE 10 *Particle-size determinations of sand (s), silt (si), and clay (c) by a class on six soil samples, as compared to the laboratory analyses*[a]

	Niagara			Canaseraga			Red Hook			Wayland			Arkport			Tioga			Percentage missed
	s	si	c	s	si	c	s	si	c	s	si	c	s	si	c	s	si	c	
Laboratory analyses	7	51	42	24	46	30	18	59	23	50	37	13	83	10	7	55	35	10	
Student 1	0	50	50	25	35	40	10	60	30	65	10	25	80	15	5	50	0	50	198
Student 2	10	10	80	45	0	55	0	40	60	40	40	20	80	10	10	60	10	30	324
Student 3	20	0	80	45	0	55	10	10	80	30	30	40	80	10	10	60	30	10	378
Student 4	15	35	55	10	45	45	10	60	30	57	18	25	80	10	10	30	60	10	177
Student 5	10	50	40	30	35	35	55	5	40	70	15	15	90	5	5	60	20	20	230
Student 6	20	0	80	45	0	55	20	10	70	30	30	40	80	10	10	60	10	30	402
Student 7	20	70	10	10	50	40	10	58	32	55	25	20	88	9	3	67	16	17	182
Student 8	15	35	50	10	45	45	10	60	30	57	18	25	60	35	5	30	60	10	216
Student 9	50	40	10	30	40	30	10	60	30	60	30	10	85	10	5	65	25	10	158
Student 10	20	40	40	40	0	60	0	40	60	60	20	20	0	10	90	70	10	20	442
Student 11	5	20	75	15	25	60	25	30	45	35	30	35	55	35	10	45	20	35	334
Student 12	20	30	50	18	42	40	20	40	40	50	20	30	90	9	1	75	10	15	197
Student 13	10	45	45	10	30	60	10	60	30	80	5	15	90	5	5	60	15	25	206
Student 14	0	50	50	25	30	45	10	60	30	65	10	25	80	15	5	60	0	40	198
Student 15	10	30	60	20	40	40	10	60	30	50	30	20	85	10	5	60	10	30	146
Student 16	20	40	40	20	35	45	10	65	25	30	40	30	45	40	15	45	20	35	228
Student 17	15	30	55	15	35	50	15	50	35	57	18	25	20	70	10	45	45	10	290
Percentage missed	173	329	296	181	303	290	167	261	306	211	252	205	289	162	133	177	354	217	4,306
Average	10	19	17	11	18	17	10	15	18	12	15	12	17	10	8	10	21	13	

[a]The "percentage missed" values were obtained by subtracting the laboratory number and the field determination for each student and each percentage estimate. A low "percentage missed" value is a good score; the high values indicate larger deviations of the field estimate from the laboratory data. The average numbers are the averages of the determinations of all the 17 students, and can be compared with each of the laboratory analyses for each of the samples.

25

was more difficult for the class to estimate in most soils, because silt can be confused with clay content until experience helps an individual to determine proportions from relative amounts of smoothness and stickiness in the fine earth fractions of soil samples. On the average, however, the class did well in estimating sand, silt, and clay in the samples. With more experience, individuals could easily "fine-tune" their fingers to more closely correlate their estimates with the laboratory data.

The samples reported in Table 10 were collected from contrasting soil map units on the campus of Cornell University (Neeley et al., 1965), but similar contrasting samples can also be collected from soil map areas almost anywhere for teaching and learning about determination of soil texture. If laboratory facilities are not available, then percentage proportions can be estimated from the defined soils of the map units; soil scientists and conservationists familiar with the Cooperative Soil Survey can also provide assistance in teaching and learning about soil texture and other soil properties. Exercises such as that illustrated by the data in Table 10 demonstrate the precision by which soil properties can be described in the field, and show clearly the value of the soil surveys and soil information for interpretive purposes.

General Classes

Sometimes it is convenient to group soil textural classes into more general categories, as illustrated in Table 9. Soils may be termed sandy, loamy, or clayey or called coarse, medium, and fine textured. Coarse fragments (Table 9) are described as modifiers of soil texture names according to their shape and volume in the soil mass (Fig. 10 can be used for estimating volumes of coarse fragments in a pit or on the soil surface). Terms such as "gravelly silt loam" or "channery silt loam" are used when the soil mass contains about 20-50 percent coarse fragments by volume. When coarse fragments comprise 50-90 percent of the soil volume terms like "very gravelly" or "very channery" or "very stony" are used. Stoniness and rockiness classes (Table 9) can be used in relation to tillage of the soil surface, but other categories can also be used for soil excavations and for forestry. Coarse fragments, stoniness, and rock outcrops cause many problems in use of areas where they are numerous, and they are generally mapped according to the detail needed in utilizing and interpreting the soil map units.

In the family categories of Soil Taxonomy (Soil Survey Staff, 1975), all soils are grouped according to particle size and mineralogy, among other things. The particle-size groupings are intended to be important for interpretations and use of broad areas of soils (not as specific as detailed soil map units). The classes of particle sizes are given in Figure 14—notice that the boundary lines for clayey, loamy, silty, and sandy soils are different from soil textural class boundaries in Figure 11. The particle-size groups in Figure 14 are used to describe the "average" particle size of the several horizons or the particle size when soil layers are mixed in the "control section" down to a depth of about 1 meter or to the top of hard layers or bedrock.

Sandy soils (Fig. 14) include sands and loamy sands but exclude loamy very fine sand and very fine sand (Soil Survey Staff, 1975). Loamy soils include loamy very fine sand, very fine sand, and finer textures with less than 35 percent clay; loamy soils are subdivided into coarse-loamy, fine-loamy, coarse-silty, and fine-silty categories. Coarse-loamy soils have a loamy particle size that has 15 percent or more by weight of fine sand (0.25–0.1 mm) or coarser particles (up to 7.5 cm), with less than 18 percent clay in the fine earth fraction. Fine-loamy soils have 15 percent or more fine sand (0.25–0.1 mm) or

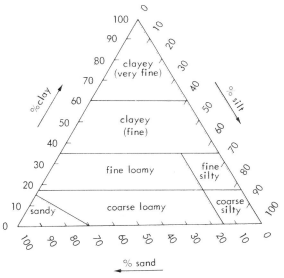

FIGURE 14 *Weight percentages of sand, silt, and clay for classification in soil families (adapted from Soil Survey Staff, 1975).*

coarser particles (up to 7.5 cm) with 18-35 percent clay in the fine earth fraction. Coarse-silty soils have less than 15 percent of fine sand (0.25-0.1 mm) or coarser particles (up to 7.5 cm), and less than 18 percent clay in the fine earth fraction. Fine-silty soils have less than 15 percent of fine sand (0.25-0.1 mm) or coarser particles (to 7.5 cm), with 18-35 percent clay in the fine earth fraction. Clayey soils (fine and very fine) have 35 percent or more clay by weight (Fig. 14) and less than 35 percent coarse fragments by volume. Fine soils have a clayey particle size that has 35-60 percent clay in the fine earth ($<$ 2 mm diameter) fraction; very-fine soils have 60 percent or more clay in the fine earth. If coarse fragments in the control section of soils are greater than 35 percent by volume, then particle-size groupings into soil families are termed "skeletal," such as sandy-skeletal, loamy-skeletal, and clayey-skeletal.

STRUCTURE

Structure is the aggregation of primary soil particles into compound particles (clusters of primary particles) which are separated from adjoining aggregates by surfaces of weakness (Soil Survey Staff, 1962, 1975). Structure is important because it is a reflection of the genetic processes that formed the soil, and in turn structure helps to determine infiltration rates, shrink-swell, and other aspects of soil behavior under crop management systems and engineering performance stresses. Structure is related to texture and other soil properties, so that a description of structure gives information about use of soil areas beyond the mere measure of aggregation units. Older soils more highly weathered generally have better developed structure than do younger soils in sediments where much of the structure may be inherited from the original geologic materials. Often measurement and description of structure help in prediction of soil performance, as where platiness in soil horizons obstructs water flow through soils, or where plates inclined downslope indicate soil movement in landslides (past, present, or future). Shrinking and swelling in soils often produces angular blocky structures, with slickensides or smeared clay coats on the ped surfaces—these soils have obvious hazards for many uses.

Field descriptions of soil structure must indicate the shape and arrangement of the aggregates or peds, the size, and their distinctness or durability. Each of these three qualities requires separate sets of terms (Soil Survey Staff, 1962, 1975). Shape and arrangement of peds are designated as type of soil structure; size of peds, as class; and degree of distinctness, as grade. Between the peds or aggregates, of course, are pore spaces and films or stains on ped faces that must be described also. Ped interiors are often of different color and composition than the exteriors, so that all of the description components (horizon, color, texture, structure, consistence, reaction, boundary) must be considered to be interrelated for the proper interpretation of the soil profile description.

Soils have four primary types of structure:

1. *Spheroidal* or *polyhedral*, with particles arranged around a point and bounded by curved or very irregular surfaces that are not accommodated to the adjoining aggregates.
2. *Platy*, with particles arranged around a plane, generally horizontal.
3. *Blocklike* or *polyhedral*, with particles arranged around a point and bounded by flat or rounded surfaces which are casts of the molds formed by the faces of surrounding peds.
4. *Prismlike*, with particles arranged around a vertical line and bounded by relatively flat vertical surfaces.

Table 11 presents the types and classes of soil structure. Figures 15-18 give the relative size limits for structures, arranged in classes for easy comparisons of the size charts with peds from soils in the field. Spheroidal structure is most common in surface horizons, and may be granular (relatively nonporous) or crumb (very porous). Plates generally overlap at the edges. Blocks may be angular blocky (bounded by planes intersecting at relatively sharp angles) or subangular blocky (having mixed rounded and plane faces with vertices mostly rounded). Soil prisms may be defined as prismatic (without rounded upper ends) or columnar (with rounded caps). Many soils have compound structure, with prisms breaking into blocks of different sizes, or some soils may have horizons with both granular and blocky structures.

Grade of structure is the degree of aggregation and expresses the differential between cohesion within aggregates and adhesion between aggregates; it is determined mainly from the durability of the aggregates and the proportions of aggregated and unaggregated materials that result when the peds are displaced or gently crushed. The grade varies somewhat in wet, moist, and dry soils and should be determined at the most important moisture contents. Structures generally become stronger and more distinct when they are exposed in roadcuts, but some soil horizons are unstable due to the breakdown of soil structures upon weathering in roadcuts. The terms for describing grade of structure are structureless, weak, moderate, and strong.

Structureless is that condition (Soil Survey Staff, 1962, 1975) with no observable aggregation or no definite orderly arrangement of natural lines of weakness. If the volume of the soil horizon is coherent, the term "massive" is used; if the horizon consists of single grains (as for loose sand) it is called "single grain."

TABLE 11 *Types and classes of soil structure (adapted from Soil Survey Staff, 1962, 1975, and Olson, 1976)*

Class	Type (Shape and arrangement of peds)						
	Platelike with one dimension (the vertical) limited and greatly less than the other two; arranged around a horizontal plane; faces mostly horizontal.	Prismlike with two dimensions (the horizontal) limited and considerably less than the vertical; arranged around a vertical line; vertical faces well defined; vertices angular.		Blocklike, polyhedronlike, or spheroidal, with three dimensions of the same order of magnitude, arranged around a point.			
				Blocklike; blocks or polyhedrons having plane or curved surfaces that are casts of the molds formed by the faces of the surrounding peds.		Spheroids or polyhedrons having plane or curved surfaces which have slight or no accommodation to the faces of surrounding peds.	
		Without rounded caps	With rounded caps	Faces flattened; most vertices sharply angular	Mixed rounded and flattened faces with many rounded vertices	Relatively non-porous peds	Porous peds
	Platy	Prismatic	Columnar	Angular blocky	Subangular blocky	Granular	Crumb
Very fine or very thin	Very thin platy; thinner than 1 mm	Very fine prismatic; diameter less than 10 mm	Very fine columnar; diameter less than 10 mm	Very fine angular blocky; less than 5 mm	Very fine subangular blocky; less than 5 mm	Very fine granular; less than 1 mm	Very fine crumb; less than 1 mm
Fine or thin	Thin platy; 1-2 mm	Fine prismatic; 10-20 mm	Fine columnar; 10-20 mm	Fine angular blocky; 5-10 mm	Fine subangular blocky; 5-10 mm	Fine granular; 1-2 mm	Fine crumb; 1-2 mm
Medium	Medium platy; 2-5 mm	Medium prismatic; 20-50 mm	Medium columnar; 20-50 mm	Medium angular blocky; 10-20 mm	Medium subangular blocky; 10-20 mm	Medium granular; 2-5 mm	Medium crumb; 2-5 mm
Coarse or thick	Thick platy; 5-10 mm	Coarse prismatic; 50-100 mm	Coarse columnar; 50-100 mm	Coarse angular blocky; 20-50 mm	Coarse subangular blocky; 20-50 mm	Coarse granular; 5-10 mm	Coarse crumb; 5-10 mm
Very coarse or very thick	Very thick platy; thicker than 10 mm	Very coarse prismatic; diameter more than 100 mm	Very coarse columnar; diameter more than 100 mm	Very coarse angular blocky; more than 50 mm	Very coarse subangular blocky; more than 50 mm	Very coarse granular; more than 10 mm	Very coarse crumb; more than 10 mm

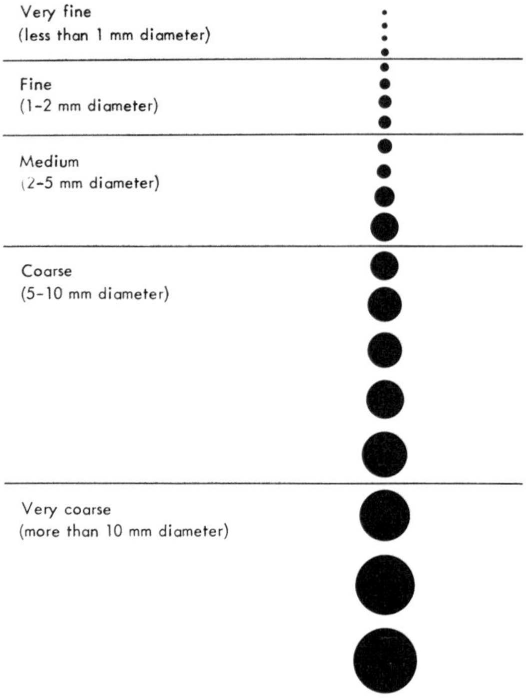

Very fine
(less than 1 mm diameter)

Fine
(1–2 mm diameter)

Medium
(2–5 mm diameter)

Coarse
(5–10 mm diameter)

Very coarse
(more than 10 mm diameter)

FIGURE 15 *Relative size limits for granular and crumb structures (adapted from Soil Conservation Service field notebook pages).*

Very fine
(less than 1 mm thick)

Fine
(1–2 mm thick)

Medium
(2–5 mm thick)

Coarse
(5–10 mm thick)

Very coarse
(more than 10 mm thick)

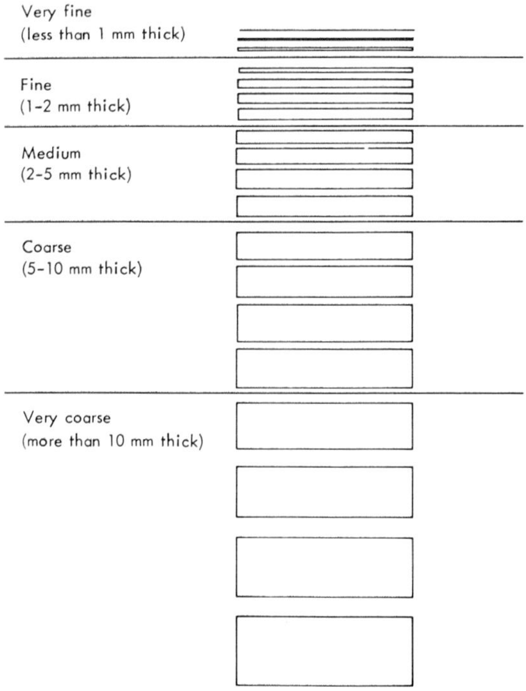

FIGURE 16 *Relative size limits for platy structures (adapted from Soil Conservation Service field notebook pages).*

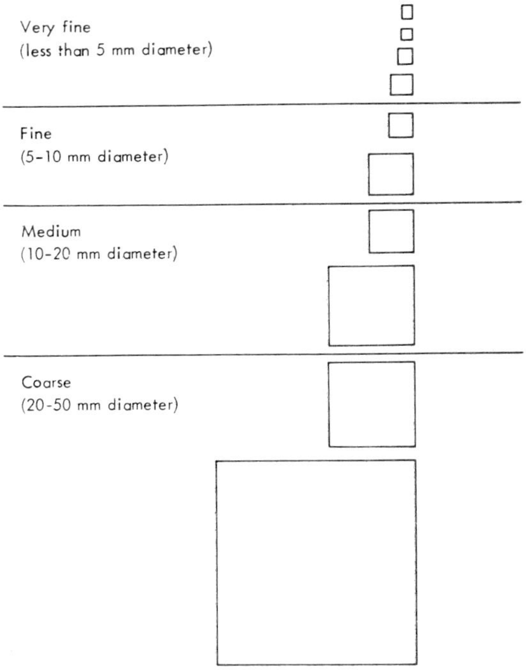

Very fine
(less than 5 mm diameter)

Fine
(5–10 mm diameter)

Medium
(10–20 mm diameter)

Coarse
(20–50 mm diameter)

FIGURE 17 *Relative size limits for angular and subangular blocky structures (adapted from Soil Conservation Service field notebook pages).*

Very fine
(less than 10 mm diameter)

Fine
(10–20 mm diameter)

Medium
(20–50 mm diameter)

Coarse
(50–100 mm diameter)

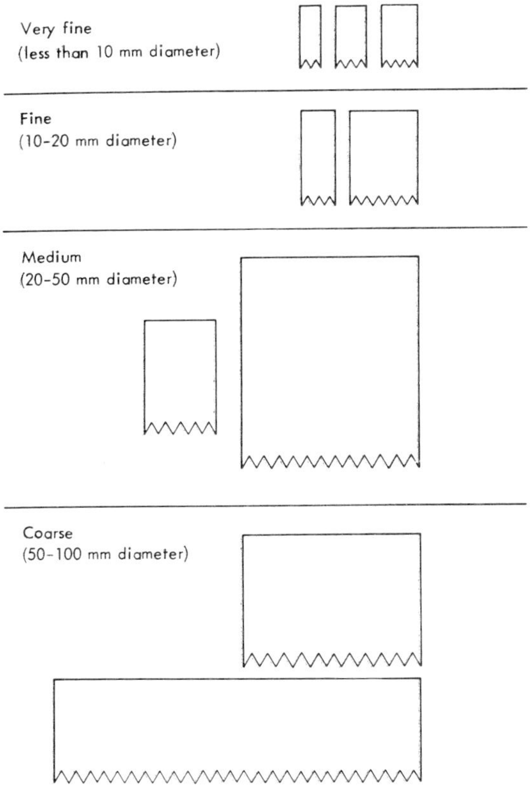

FIGURE 18 *Relative size limits for prismatic and columnar structures (adapted from Soil Conservation Service field notebook pages).*

FIGURE 19 *Weak fine granular and weak medium subangular blocky structures from the B horizon of an Oxisol in Brazil. Local farmers call this soil material "coffee dust" because it is so permeable, well drained, and light and fluffy.*

FIGURE 20 *Moderate to strong fine platy and very fine angular blocky structures from the C horizon of an Entisol in Turkey. This structure was primarily formed by sliding movements in the soil mass of a recent landslide.*

Weak structure is that degree of aggregation characterized by poorly formed indistinct peds that are barely observable in place. Disturbed soil material with weak structure breaks into a mixture of a few entire peds, many broken peds, and much disaggregated material. The terms "very weak" and "moderately weak" may be used, if necessary.

Moderate grade of structure has well-formed distinct peds that are moderately durable and evident—but not distinct in undisturbed soil. Disturbed soil horizons of this grade break down into a mixture of many distinct entire peds, some broken peds, and a little disaggregated material.

Strong structure has durable peds that are quite evident in undisturbed soil, that adhere weakly to one another, and that withstand displacement and become separated when the soil is disturbed. When disturbed from soil horizons, strong structure consists very largely of entire peds with a few broken peds and little or no disaggregated material. Terms "moderately strong" and "very strong" may be used, if necessary.

Structures are named in sequence of (1) grade (distinctness), (2) class (size), and (3) type (shape). Thus, peds loosely packed and roundish but not very porous, mostly between 1 and 2 mm in diameter, and quite distinct would be classified as having strong fine granular structure. Figure 19 shows an example of weak fine granular and weak medium subangular blocky structures. Figure 20 illustrates moderate to strong fine platy and very fine angular blocky structures. Figure 21 shows a soil with moderate coarse platy structure. Figure 22 is a soil with strong coarse to fine subangular blocky structure. Each soil in a distinctive place in the landscape has its own particular soil structure and other characteristics, which determine how it is classified and how it can best be used in each landscape position.

FIGURE 21 *Moderate coarse platy structure from the B horizon of an Alfisol in Wisconsin formed in glacial till. Water slowly percolating through this soil is forced to travel extra distances around the plates in a complex series of cracks and pores.*

FIGURE 22 *Strong coarse to fine subangular blocky structure from the B horizon of an Ultisol in Brazil. This soil is moderately weathered and the structure is well developed.*

CONSISTENCE

Soil consistence is described according to the degree and kind of cohesion and adhesion and by the resistance to deformation and rupture (Soil Survey Staff, 1962, 1975). Table 12 lists the criteria for determination of soil consistence in the field. The properties and behavior of soils change with increasing water content. Description of consistence helps to enable prediction of soil behavior under stress. Soil consistence terms and other tests enable soils to be characterized according to their durability and workability in place, and their trafficability and support when manipulated and when supporting a load.

Consistence is described under dry, moist, and wet conditions. Soils of arid regions are commonly hard or very hard when dry (Table 12). Descriptive terms for dry soils include extremely hard, very hard, hard, slightly hard, soft, and loose. Extremely hard dry soils are resistant to pressure and cannot be broken in the hands; commonly a rock hammer is required for breaking horizons of these soils. Very hard dry soils are very resistant to pressure and can be broken in the hands only with difficulty; the thumb and forefinger cannot break fragments of very hard dry horizons. Hard dry soils are moderately resistant to pressure, can be broken in the hands, but commonly cannot be broken between thumb and forefinger. Hard, very hard, and extremely hard dry soils commonly have a crusting and cementing tendency in their topsoils (and low organic-matter content), so that seedling emergence is a problem. Slightly hard dry soils are only weakly resistant to pressure and can easily be broken between thumb and forefinger. Soft dry soils are only very weakly coherent and break to powder or individual grains under slight pressure. Loose dry soils are noncoherent. Loose and soft dry soils are commonly sandy, and lack enough clay or fines for coherence of the peds or aggregates.

TABLE 12 *Field criteria for determination of soil consistence (adapted from Soil Survey Staff, 1962, 1975, and Olson, 1976)*[a]

Dry Consistence—Air dry

   Loose—falls apart without handling. Cannot pick up a ped.
   Soft—can be picked up as a mass but falls apart with slight pressure and barely indents the fingers.
   Slightly hard—a ped or clod can be picked up. Before breaking between thumb and forefinger it indents the finger deeply but breaks without strong pressure.
   Hard—must exert strong pressure to break. Can be broken between thumb and forefinger under strongest pressure one can exert.
   Very hard—cannot be broken between thumb and forefinger.

Moist Consistence—About halfway between air dry and field capacity

   Loose—falls apart without handling. Cannot pick up a ped.
   Very friable—crushes with only slight indentation of finger.
   Friable—indents finger when crushed but only gentle pressure is needed.
   Firm—crushes only when deliberate pressure is applied. Deeply indents the fingers.
   Very firm—can barely be crushed between thumb and forefinger.
   Extremely firm—cannot be crushed between thumb and forefinger.

Wet Consistence—Slightly above field capacity

   Stickiness—press between thumb and finger
   Nonsticky—almost none adheres to either finger
   Slightly sticky—adheres to both fingers but finally pulls cleanly free of one without stretching
   Sticky—stretches noticeably before breaking and leaves material on both fingers
   Very sticky—stretches as one exerts strong effort to pull fingers apart
   Plasticity—Roll and deform
   Nonplastic—cannot form a wire by rolling in fingers.
   Slightly plastic—can form a wire by rolling. The wire will not support its own weight. Easily deformed under pressure.
   Plastic—can form a wire that will bear its own weight. Must press to deform.
   Very plastic—can form a strong wire that will whip. Must exert strong pressure to deform.

[a]Consistence expresses degree and kind of adhesion and cohesion, and is given for dry, moist, and wet soil.

Moist consistence is determined in the field a few days after a rain, when the soils are between the air-dry state and field capacity. Dry soils may be moistened artificially to measure moist consistence. Loose moist soil is noncoherent. Very friable moist soil crushes under gentle pressure but coheres when pressed together in a mass. Friable moist soil crushes easily under gentle to moderate pressure between thumb and forefinger, and coheres when pressed together. Firm moist soil can be crushed under moderate pressure between thumb and forefinger but noticeable resistance is felt. Very firm moist soil can be crushed under strong pressure in the hands, but usually not between thumb and forefinger. Extremely firm moist soil cannot be crushed between thumb and forefinger and must be broken apart bit by bit.

Wet consistence is determined when soils are saturated with water. Both stickiness and plasticity must be measured. Stickiness is the adhesion of the soil to other objects, and is observed by pressing wet soil between thumb and forefinger. Nonsticky wet soil does not adhere to thumb or finger. Slightly sticky wet soil adheres only slightly and is not stretched when the digits are separated. Sticky wet soils stick to the fingers and stretch when the fingers are pulled apart. Very sticky wet soils adhere strongly to the fingers and stretch considerably when the fingers are separated. When tractors attempt to traverse wet sticky soils, the soil materials stick to the tractor tires as to the fingers in the wet consistence test.

Plasticity of wet soils is the ability to change shape continuously under the influence of an applied stress and to ⬧tain the impressed shape on removal of the stress. In the field, wet soil samples are rolled into a wire between thumb and fingers. Wet soil is non-plastic if no wire is formable. Wet soil is slightly plastic if a wire if formable but the soil mass is easy to deform. Plastic wet soil can form into a wire but pressure is required for deformation. Very plastic wet soils can be formed into a strong wire that will whip, and strong pressure is required for deformation. Tractor tires and automobiles are easily bogged down in wet very plastic soils, and vehicles are difficult to extract from mudholes in these soils.

Soil consistence is sometimes altered by cementation, which may be irreversible. Cementation refers to brittle hard consistence caused by calcium carbonate, silica, iron, aluminum, or other materials other than clay minerals. Usually, the cementation is not altered by moistening, but groundwater generally had an influence in the original formation of the cementation. Figure 23 illustrates a coarse fragment of ironstone (hardened plinthite). Cementation by carbonates and silica are also common in drier regions. Weakly cemented materials are brittle and hard but can be broken in the hands. Strongly cemented soil horizons can be broken in the hand but are more easily broken with a ham-

FIGURE 23 *Fragment of ironstone (hardened plinthite) gravel taken from a borrow pit at the edge of the peneplain of the central plateau of Brazil. It is about 10 cm across and the moist colors range from reddish black (10R 2/1) to brownish yellow (10YR 6/6). In a previous geologic period, groundwater moved iron into the low-lying strata, which hardened, and now the resistant ironstone caps the plateau. The strata once low in the landscape became hard and resistant to erosion through cementation.*

mer. Indurated soil materials are very strongly cemented. They can be broken with hammers or heavy crushing equipment and used as aggregate material for road construction and other engineering works.

REACTION

The reaction of soil samples with chemical indicator solutions in the field provides excellent techniques to quickly obtain information about the chemical content of the soil horizons. These tests can be conveniently used during soil mapping while the soil scientist is digging in the soils and examining them, or can be easily used by others looking at soils. The field test for pH is perhaps the most valuable—it is simple, quick, and reliable. The pH of a soil indicates a great deal about the nutrient status and the weathering processes that have formed the soil. The field pH test can also be used directly to obtain preliminary recommendations about liming and fertilizing when more detailed soil laboratory data are not immediately available.

The degree of soil acidity or alkalinity (Soil Survey Staff, 1962) is expressed by pH—the logarithm of the reciprocal of the H^+ ion concentration. In the pH notation, pH 7 is neutral; lower values indicate acidity, and higher values show alkalinity. Soil horizons vary in pH from about 3 to about 10. Soils of humid regions that are highly weathered and leached are commonly acid; soils of arid regions often have high pH where salts and bases have accumulated in the landscape as seepage waters have evaporated. The terms used to describe pH ranges in soils are:

Term	pH
Extremely acid	< 4.4
Very strongly acid	4.5-5.0
Strongly acid	5.1-5.5
Medium acid	5.6-6.0
Slightly acid	6.1-6.5
Neutral	6.6-7.3
Mildly alkaline	7.4-7.8
Moderately alkaline	7.9-8.4
Strongly alkaline	8.5-9.0
Very strongly alkaline	> 9.1

In a state or region, pH test kits are commonly made to test the range of soils in a given area (Fig. 7). The kits are then used by soil scientists, farmers, environmentalists, foresters, geologists, land economists, tax appraisers, and many others. Indicator dyes are added to small samples of soil on a spot plate. The reagent color, when reacted with soil, quickly develops and is compared with standard color chips on color charts calibrated to pH values (much as a Munsell Soil Color Chart is used to describe soil colors). These colorimetric methods usually give satisfactory results on mineral soils between about pH 3.8 and pH 9.6. A soil scientist or experienced soil tester should get quick pH test results within a few minutes of 0.2 to 0.4 unit of the pH values that would be determined from the same soil samples electrometrically in the laboratory. Common specific indicators used in the field and their approximate pH ranges are:

Indicator	pH range
Bromcresol green	3.8-5.6
Chlorophenol red	5.0-6.2
Bromthymol blue	6.0-7.2
Phenol red	6.8-8.4
Cresol red	7.2-8.8
Thymol blue	8.0-9.6

The pH test kit illustrated in Figure 7, for example, has chlorophenol red and bromthymol blue indicators with ranges of about 5.0-6.2 and 6.0-7.2, respectively. Together in a kit the indicators can test soil in a region or state from about 5.0 to 7.2—from very strongly acid to neutral. The overlap in the ranges permits a constant check on the determinations in the middle part of the range when both indicators are used on separate samples of the same soil horizon.

In a practical sense, for example, the pH test can be used by a farmer to determine the approximate amount of lime application needed to increase the soil pH for various crops. Table 13 illustrates the tons of lime needed per acre to raise the soil pH to 7.0 for alfalfa in New York State in different soil textures. The soil texture groupings are essentially a grouping of soil profiles or soil map units. The actual field situation is much more complicated than Table 13 indicates, because of other factors, such as soil wetness and subsoil differences. The soil survey and soil profile descriptions must be integrated into the lime and fertilizer recommendations for the soil test results to be most reliable. The chemical test and physical soil survey are complementary: the pH test helps describe the soil profile, and the soil profile description assists in interpreting the soil test significance under actual field conditions. The reactivity of lime added to the soil is much dependent upon the fineness to which the limestone is ground. The larger amounts of lime listed in

TABLE 13 *Amount of lime application (tons per acre for plow layer) required to increase the soil pH to 7.0 for alfalfa in well-drained soils of different textures and relatively uniform profiles*

Initial pH from soil test kit	Surface textures of different soil profiles			
	Sand	Sandy loam	Loam and silt loam	Silty clay loam and clay
< 4.5	4.0	7.0	11.0	15.0
4.6-4.7	3.5	6.5	10.0	13.0
4.8-4.9	3.0	6.0	9.0	12.5
5.0-5.1	2.5	5.5	8.5	12.0
5.2-5.3	2.0	4.5	6.5	8.0
5.4-5.5	1.5	3.0	4.0	6.0
5.6-5.7	1.0	2.5	3.0	5.5
5.8-5.9	1.0	2.0	2.5	3.5
6.0-6.1	1.0	1.5	2.0	3.0
6.2-6.3	0.5	1.0	1.5	2.5
6.4-6.5	0.5	1.0	1.5	2.0
6.6-6.7	0.5	1.0	1.0	1.5

Table 13 need to be added in split amounts over periods of several years to be most effective.

Many other chemical tests can also be used in the field, to test for other minerals and compounds in the soils. The presence of free carbonates in the soil, for example, may be tested with 10 percent hydrochloric acid. Effervescence (gas given off) indicates CO_2 in the reaction:

$$CaCO_3 + 2HCl \rightarrow CO_2 \uparrow + CaCl_2 + H_2O$$

The presence of free carbonates as indicated by the CO_2 bubbles indicates that the soil has been very little leached, has had relatively little weathering, or has had $CaCO_3$ accumulations or depositions in or on the soil. In some soils of humid regions, carbonates have been leached out of the A and B horizons and accumulated in the C horizon. If the soil parent material was originally calcareous, then the presence of free carbonates (as indicated by the HCl acid test) is a good indication of the boundary line between the B and C horizons in some soils. Soil profiles that have calcareous subsoils, of course, need less lime to be added for crops than do soils with acid subsoils, because the plant roots from crops such as alfalfa can exploit the subsoil lime for crop growth. Thus, interpretation of any test of the surface soil must also consider the entire soil profile and its characteristics that influence root nutrition and crop response.

BOUNDARY

The boundary of soil horizons describes the nature of the transition from one horizon to another. The boundaries generally are closely related to the genesis of the soil. Thus, horizons of soils formed in alluvial (floodwater deposited) and lacustrine sediments are often layered, and in places the strata are highly contrasting and the boundaries may be abrupt from one horizon to another. In other places soils have formed in materials that are uniform, or are mixed, or are influenced by shrinking and swelling and soil movement. Boundaries of horizons in these soils, in contrast, may be vague or diffuse.

Soil horizon boundaries (Soil Survey Staff, 1962) are described according to (1) distinctness and (2) surface topography. Boundary distinctness is sharp and readily apparent in some soils and diffuse and vague in others where one horizon gradually merges into another. In diffuse horizons, the location of the boundary requires careful comparisons of small samples of soil from different parts of the profile until the midpoints and transition areas of the horizons are established. Small markers are commonly inserted into the soil profile until all horizons are designated, then measurements are taken, and finally the individual horizons can be described and sampled.

The distinctness of soil horizons depends partly upon the contrast between them and partly upon the width of the boundary itself or the amount of the profile in the transition between one horizon and the next. Many adjacent horizons are highly contrasting in several features, so that the boundaries can be delineated relatively easily. The characteristic widths of boundaries between soil horizons are described as:

1. Abrupt—if less than 1 in. wide
2. Clear—if about 1-2½ in. wide
3. Gradual—if 2½-5 in. wide
4. Diffuse—if more than 5 in. wide

Soil profiles, of course, are not two-dimensional units because they represent considerable areas of soil map units. Observations of soil horizons are made in profiles or sections, and are so photographed and described, but horizons are not bands or literal horizons but rather three-dimensional parts of pedons (volumes of soil) that may be smooth or exceedingly irregular. Some soils have horizons with lower boundaries that are a smooth surface; other soils have horizons with wavy, irregular, or broken boundaries. The boundaries of soil horizons are described as:

1. Smooth—if nearly a plane
2. Wavy—if pockets are wider than their depth
3. Irregular—if pockets are deeper than their width
4. Broken—if parts of the horizon are unconnected with other parts

The lower boundary of each horizon in a soil profile description is conventionally stated as part of each horizon description. The deepest lower boundary in a soil profile is not given if it is not seen and described; the lowest horizon boundary is usually described as continuing to greater depth in the measurements giving the depth and thickness of that horizon in the soil profile description. The notation "28-36+" in Table 3, for example, indicates that the soil scientist describing that soil dug into the C horizon from 28 in. to 36 in., but did not go deeper than that. That C horizon appeared to continue, but the describer did not see the bottom soil boundary.

CHAPTER 3

LABORATORY ANALYSES

Although soil profile descriptions provide the most valuable data about soils, laboratory analyses also constitute an important data base to supplement the field descriptions. The field descriptions basically establish the soil in the context of the landscape, and the laboratory data give additional information in precise numerical format about the chemical, physical, and behavioral aspects of the soils. The purpose of all soil data, in the ultimate sense, is to enable performance to be predicted on different soil landscape units (soil map units). Actual performance data must be correlated with laboratory analyses and soil profile descriptions for the system to be most effective. Thus, soil lab tests are correlated with greenhouse pot tests of plant growth on various soils, with experiment station results of different fertilizer applications on specific soils, and with test trials on soils in different farmers' fields. Correlations of lab results with yields are also a test of the usefulness of the soil description and mapping effort. Similarly, engineering lab tests are correlated with roadbed performance and foundation failure on different soils. Soil characterization data are being used increasingly to classify soils in the Soil Taxonomy System (Soil Survey Staff, 1975), and increasingly, numerical data about soils are being used in computer manipulations and predictions.

Basically, three major kinds of laboratory analyses are most commonly used on samples collected from soil map units. These analyses are:

1. Soil fertility laboratory tests
2. Engineering soil tests
3. Soil classification lab analyses

SOIL FERTILITY TESTS

Soil fertility tests are "quick tests" used to analyze topsoil samples from farmers' fields to determine "available" nutrients and make recommendations for fertilizer applications. Usually, samples are collected and analyzed every year or every few years, and recommendations are based on the nutrient needs of the crop to be grown. Samples are usually collected from a number of small cores taken from a single soil map unit within a farmer's field, and composited (mixed together) to constitute a representative soil sample. The sample is then air-dried and submitted to laboratory testing. Many different procedures are used by different laboratories to achieve correlations between the nutrient content in the extracting solution and crop growth in the field; the New York State laboratory (Greweling and Peech, 1965), for example, uses a sodium acetate–acetic acid extract. The extracting solution is leached through the soil sample, and the leachate solution is then used for the determinations of amounts of the various available nutrients.

The chemical analyses of topsoil samples, of course, determine only the content of the nutrients tested; they do not indicate the subsoil conditions. Thus, in interpretation of the chemical values, it is essential that the lab test results be related to the soil profiles in which the crop is to be grown. A wet soil, for example, might have high values from the chemical soil test for growing alfalfa—but high water tables and frost heaving would kill the alfalfa plants. Similarly, a fragipan (dense impermeable subsoil layer) is not indicated by the chemical soil test, but it hinders root development due to the poor physical conditions. Increasingly, soil test recommendations are being made by correlation to specific soil profiles and soil map units, and computer systems can develop the correlations and recommendations. Soil maps assist in the soil sampling operations, so that each soil map unit can be recorded for each soil sample as it is collected.

Results

Table 14 illustrates typical results from soil fertility tests. The data in Table 14 were from samples collected from Seneca County in New York State in Ontario, Arkport, Schoharie, deep muck, and Lakemont soils with very different soil profiles. Ontario is a fertile responsive soil developed in glacial till, Arkport is a sandy infertile droughty soil, Schoharie is a well-drained clayey soil, the deep muck is organic material, and Lakemont is a poorly drained clayey soil. The samples illustrated in Table 14 were collected from representative areas of the different soil map units. Variability within each soil map unit is great (Beckett and Webster, 1971), but the soil test values are easily and quickly changed and influenced by lime and fertilizer applications and by farmer practices. Thus, the area represented by the ArB Arkport Sample 8 with low organic matter (0.9 percent) could have the organic matter content raised within several years if more crop residues were added to the soils. Liming would raise the pH of the more acid soils and increase the amount of calcium (Ca) in the soils. The values of many of the units in Table 14 are given in pounds per acre plow layer (parts of nutrient per 2,000,000 parts of topsoil) in order to facilitate calculations for rates of fertilizer applications. Soil tests are summarized periodically to indicate fertility levels of different soils and effects of farmer practices (Klausner and Reid, 1979).

Land-Use Effects

Some important soil profile and land-use considerations for cropping are illustrated in Table 15. These are field observations made at the time the soil samples listed in Table 14 were collected. The land use is closely correlated with the soil conditions. In early May (Table 15) the fertile Ontario soils were mostly already plowed and seeded. The infertile sandy Arkport soils, in contrast, were mostly unplowed. Schoharie soils had variable stages of land preparation due to the clayey textures. Muck soil had forest or intensive crops, depending upon the investment conditions of clearing and drainage. Wet clayey Lakemont soils were largely abandoned, or still unplowed in late June due to the wet soil conditions. The correlation of soil profiles with soil test is extremely important, and the physical conditions of wetness and texture are fully as important to land use as the fertility status of the soils.

Figure 24 shows another example of the correlation between soil chemical and soil

TABLE 14 *Soil fertility analyses of samples from selected soil map units in Seneca County, New York (analyses by Cornell Soil Test Lab)*

Sample no.	Soil map unit	Organic matter (%)	pH	Exch. H (meq/100 g)	lb/acre plow layer (2,000,000 lb)										Soluble salts: K × 10
					P	K	Mg	Ca	Mn	Fe	Al	No$_3$-N	NH$_3$-N	Zn	
1	OnB	3.0	6.3	7	2	80	125	2,450	65	7	55	41	6	<1	16
2	OnB	2.9	6.8	3	10	130	280	2,750	32	5	25	6	3	<1	12
3	OnB	3.6	7.1	4	10	140	425	3,700	67	2	15	5	1	1	12
4	OnB	4.3	6.7	7	5	110	625	4,000	111	2	20	8	3	1	12
5	OnB	4.1	6.4	7	44	290	400	3,650	73	3	10	28	17	1	25
6	OnB	2.8	6.7	6	3	95	425	2,200	55	2	25	13	3	1	10
46	OnB	2.2	7.3	3	85	190	205	2,650	48	2	15	5	2	1	10
Ave.		3.3	6.8	5	23	148	355	3,057	64	3	24	15	5	1	14
7	ArB	2.7	6.0	8	1	90	55	550	6	4	235	3	3	1	<5
8	ArB	0.9	5.8	4	15	60	20	200	4	4	100	2	1	1	<5
9	ArB	2.1	6.0	6	4	60	80	900	31	6	90	11	5	1	<5
10	ArB	1.4	5.9	7	20	100	40	350	8	8	105	2	2	2	<5
11	ArB	1.4	5.6	6	14	180	35	300	36	2	100	5	3		<5
12	ArB	1.7	6.3	4	3	90	340	1,450	55	5	20	35	1	<1	14
Ave.		1.7	5.9	6	10	97	95	625	23	5	108	10	2	1	6
13	ShB	5.1	6.1	8	2	240	400	3,250	44	3	30	7	5	1	14
14	ShB	2.1	6.8	5	14	150	525	3,400	82	3	20	9	3	1	14
15	ShB	2.8	5.8	12	3	220	550	2,750	111	42	100	5	1	1	<5
16	ShB	2.8	6.2	7	1	80	450	3,150	46	10	40	7	5	1	8
17	ShB	2.4	6.4	5	8	150	400	3,150	94	9	20	56	1	1	20
18	ShB	2.4	6.0	8	2	110	425	2,950	92	8	60	37	3	1	19
Ave.		2.9	6.2	8	5	158	458	3,108	78	12	45	20	3	1	13
19	Mr	77.0	5.2	55	20	200	17,500	19,000	42	38	55	108	24	50	80
47	Mr	51.0	7.5	11	160	680	3,050	80,000	114	4	<5	100	16	10	68
Ave.		64.0	6.4	33	90	440	10,275	49,500	78	21	30	104	20	30	74
20	LcA	5.6	5.7	10	24	125	375	3,150	29	18	25	5	5	7	14
21	LcA	7.6	7.2	5	3	125	1,325	7,250	28	4	10	8	2	1	22
22	LcA	8.2	6.8	8	5	80	900	6,750	24	6	15	7	3	1	15
23	LcA	6.5	6.9	8	6	160	1,625	7,000	78	9	20	9	4	2	28
24	LcA	9.0	6.5	12	7	150	1,125	6,500	31	36	35	10	4	2	15
25	LcA	13.0	6.6	14	8	120	1,650	8,250	18	10	30	10	6	2	20
Ave.		8.3	6.6	10	9	127	1,167	6,483	35	14	22	8	4	2	19

43

TABLE 15 *Soil and land-use conditions at sampled places in selected soil map units in Seneca County, New York*

Sample no.	Soil	Map sheet no.	Date of sampling	Land-use conditions at time of sampling
1	Ontario—OnB	7	7 May 1974	Good new seedbed with sprouts coming up; no remnants of former crops
2	Ontario—OnB	3	7 May 1974	Good alfalfa stand with grass understory on long, narrow drumlin
3	Ontario—OnB	4	7 May 1974	Small grain 6 in. tall in good stand
4	Ontario—OnB	3	7 May 1974	Recently plowed grass field
5	Ontario—OnB	6	7 May 1974	Field newly sowed to small grain
6	Ontario—OnB	2	7 May 1974	Newly plowed field
46	Ontario—OnB	4	9 Aug. 1974	Good stand of corn 6 ft. high with corn tasseling and ears filling good
7	Arkport—ArB	12	7 May 1974	Unplowed field with weeds coming up; field plowed last year, however
8	Arkport—ArB	9	7 May 1974	Unplowed field with grass and brambles coming up; field plowed a couple of years ago, however
9	Arkport—ArB	9	7 May 1974	Field newly seeded to small grain with well-decayed old cornstalks in evidence in the soil
10	Arkport—ArB	9	9 May 1974	Mowed stubble area under power line—soil probably somewhat disturbed without much A horizon
11	Arkport—ArB	6	9 May 1974	Volunteer small grain with good mulch cover in old stubble—unplowed yet this year
12	Arkport—ArB	2	9 May 1974	Newly seeded field (probably planted to corn) at edge of woods
13	Schoharie—ShB	8	9 May 1974	Stubble field being cleared of light brush cover in places—field not yet plowed this year but cropped last year
14	Schoharie—ShB	13	9 May 1974	Weedy stubble field not yet plowed this year
15	Schoharie—ShB	11	9 May 1974	Poor lawn on Eisenhower College campus, probably scalped of Ap during construction
16	Schoharie—ShB	14	9 May 1974	Stubble field recently burned—not yet plowed this year
17	Schoharie—ShB	17	9 May 1974	Field newly planted to small grain
18	Schoharie—ShB	17	25 June 1974	Good stand of newly sown small grain about 6 in. high just south of Village of Canoga
19	Deep muck—Mr	8	25 June 1974	Heavy deciduous forest in Montezuma Marsh
47	Deep muck—Mr	4	9 Aug. 1974	Good stand of potatoes with plants about 2 ft. high
20	Lakemont—LcA	11	25 June 1974	Abandoned wet farmland in weeds and brush
21	Lakemont—LcA	7	25 June 1974	Abandoned wet field growing up to weeds and brush
22	Lakemont—LcA	10	25 June 1974	Abandoned wet field with grass, weeds, and brush
23	Lakemont—LcA	7	25 June 1974	Abandoned wet field growing up to grass, weeds, and brush
24	Lakemont—LcA	7	25 June 1974	Abandoned wet field reverting to grass, weeds, and brush
25	Lakemont—LcA	7	25 June 1974	Heavy deciduous forest on wet soil with very high organic matter in soil surface under trees

physical conditions. From the pH and other soil tests, the lime requirement can be determined for different soil textures to raise the pH of the surface soil. The relationships shown in Figure 24 are the same as for those illustrated in Table 13, but the amounts are somewhat different; Figure 24 illustrates a general relationship of lime needed to raise the surface pH, and Table 13 shows the amount of lime required for the demanding crop of alfalfa. Each crop has specific requirements, and each soil profile has specific characteristics—the challenge is to match the management and the soil to produce the optimum yields. Figure 25 illustrates the correlation between the soil test for pH and the yield of a specific crop on a specific soil. Correlations are usually most effective when done on a regional or state basis. Thus, responses of crops to fertilizers on soils in New York State may be different than those in California and North Carolina and other areas.

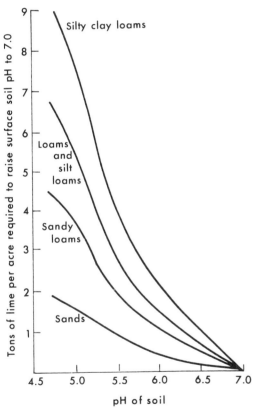

FIGURE 24   *Lime requirements of representative New York soils, showing amount of lime required to raise the soil pH to 7.0 as a function of initial pH and textural class (adapted from Lathwell and Peech, 1973).*

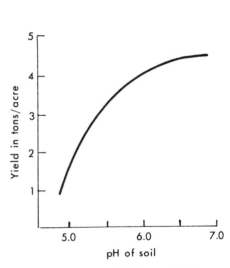

FIGURE 25   *Yield of forage (alfalfa-birdsfoot trefoil-timothy mixture) on Mardin silt loam as a function of pH of surface soil (adapted from Lathwell and Peech, 1973).*

ENGINEERING SOIL TESTS

Engineering analyses of soils (Asphalt Institute, 1969; PCA, 1973) emphasize that grain size, mineralogy, density, porosity, bearing capacity, expansion and shrinkage, cohesion, shearing, and compression are all important soil characteristics that may need to be tested for engineering purposes. Engineering analyses also indicate soil behavior under varying moisture and temperature conditions. Many engineering tests can be done on soil samples in the laboratory, and others can be performed under field conditions. Some of the procedures of the Unified soil classification system can be used as examples to illustrate some of the engineering soil tests.

Unified System

One of the most useful soil groupings for engineering purposes is the Unified soil classification system outlined in Table 16. It was developed through engineering work with many diverse soils during World War II. After World War II, it was revised and expanded in cooperation with the U.S. Bureau of Reclamation, so that it currently applies to embankments and foundations as well as to roads and airfields.

In the Unified soil classification system, all soils can be assigned to one of 15 categories based on their engineering properties (Table 16). These groups can be readily correlated with soil textures and other soil survey groupings of soil properties. Designated group symbols indicate the major characteristics. Coarse grained soils are subdivided as gravel (G) and sand (S), with four secondary divisions—GW, GP, GM, and GC (gravel); SW, SP, SM, and SC (sand)—depending on the amount and type of fines and the shape of the grain-size distribution curve. Well-graded materials (W) generally have grain-size curves that are smooth and concave, with no sizes lacking and no excess of material in

TABLE 16 *Typical names and group symbols of the Unified soil classification system (adapted from PCA, 1973)*

Group symbol	Typical names
GW	Well-graded gravel, gravel and sand mixtures, little or no fines
GP	Poorly graded gravel, gravel and sand mixtures, little or no fines
GM	Silty gravel, gravel and sand and silt mixtures
GC	Clayey gravel, gravel and sand and clay mixtures
SW	Well-graded sands, gravelly sands, little or no fines
SP	Poorly graded sands, gravelly sands, little or no fines
SM	Silty sands, sand and silt mixtures
SC	Clayey sands, sand and clay mixtures
ML	Inorganic silts and very fine sands, rock flour, silty or clayey fine sands, or clayey silts with slight plasticity
CL	Inorganic clays of low to medium plasticity, gravelly clays, sandy clays, silty clays, lean clays
OL	Organic silts and organic silty clays of low plasticity
MH	Inorganic silts, micaceous or diatomaceous fine sandy or silty soils, elastic silts
CH	Inorganic clays of high plasticity, fat clays
OH	Organic clays of medium to high plasticity, organic silts
Pt	Peat, muck, and other highly organic soils

any particular size range. Poorly graded materials (P) have excesses of materials in specific narrow size ranges. Fine-grained soils are segregated into silt (M) and clay (C), depending on their liquid limit and plasticity index. The silt and clay soil groups have secondary divisions according to their relatively low (L) or high (H) liquid limit. Silt and clay groups with high organic (O) content are designated OL or OH. Soils composed mostly of organic matter, usually very compressible and with undesirable construction characteristics, are classified as one group designated Pt.

The liquid limit of a soil is the percentage water content at which the soil passes from a plastic to a liquid state, as determined by a standard procedure (Asphalt Institute, 1969). About 50 g of soil is passed through a number 40 (40 mesh per inch) sieve and moistened with 15–20 ml of water. The wet soil is mixed to a uniform mass of stiff consistence and placed in the bottom of the curved brass cup of a "mechanical liquid limit device." The soil is spread, squeezed, and leveled in the brass cup, and separated into two parts with a "grooving tool." The cup containing the sample is lifted and dropped onto a firm base by turning the eccentric crank on the liquid limit device at a rate of 2 revolutions per second until the two sides of the soil sample come into contact at the bottom of the groove along a distance of about ½ in. The number of shocks required to close the groove is recorded. A slice of soil is taken from the brass cup, weighed, oven-dried to a constant weight at $110°C$ (for about 24 hours), and weighed again. The loss in weight due to drying is recorded as the weight of water. These operations are repeated for at least two additional portions of the soil sample in more fluid conditions. The object is to have at least one determination in the 15–25, 25–30, and 30–35 range of shocks. The water content is calculated according to the formula

percentage moisture = (weight of water/weight of oven-dry soil) $\times$ 100

To obtain the liquid limit, a flow curve is plotted on semilogarithmic graph paper (representing the relation between water content and corresponding number of shocks) with the moisture contents as abscissas on the arithmetic scale and the number of shocks as ordinates on the logarithmic scale. The flow curve should be a straight line drawn as nearly as possible through the three or more plotted points. The unit percentage moisture content corresponding to the intersection of the flow curve with the 25 shock ordinate is taken as the liquid limit of the soil.

The plastic limit of a soil is the lowest percentage water content at which the soil remains plastic, as determined by a standard procedure (Asphalt Institute, 1969). About 8 g of moistened soil is taken from the excess of the sample prepared for the liquid-limit test. The soil mass should be at a moisture content to be plastic enough to be easily shaped into a ball without sticking to the fingers excessively when squeezed. The soil mass is shaped in the hand into an ellipsoidal-shaped mass, which is then rolled between the fingers and a flat horizontal surface. A uniform thread of soil about $1/8$ in. in diameter is rolled at a rate of about 80 to 90 strokes per minute. The thread is broken, squeezed together, and rolled again until it becomes so dry that it crumbles and will not roll into a thread. At that point the plastic limit of the soil has been reached and the portions of the crumbled soil are gathered together, placed in a container, and weighed. The soil sample is oven-dried at $110°C$ to constant weight and weighed again. The loss in weight is the loss of water.

The plastic limit of the soil is expressed as the water content in percentage of the weight of oven-dry soil:

plastic limit = (weight of water/weight of oven-dry soil) $\times$ 100

Plasticity Index

The plasticity index is the difference between the liquid limit and the plastic limit, calculated by the formula

plasticity index = liquid limit − plastic limit

Some loose, coarse-textured soils, of course, are nonplastic, so that these limits cannot be determined. The Atterberg limits (liquid limit, plastic limit, plasticity index), however, enable much to be learned about engineering soil behavior with relatively simple tests. Even visual examinations of soil materials will enable a first approximation of soil placement into the Unified soil groups (Table 16). Other engineering tests determine grain size, mineralogy, shrink–swell, compressibility, compaction, and permeability. Some engineering tests require expensive equipment and elaborate facilities, but many are relatively inexpensive. Much engineering data can be interpolated from soil surveys (Soil Survey Staff, 1971; FAO, 1973; Olson and Warner, 1974), especially where some engineering tests have been conducted on soil samples collected during the course of the soil survey mapping operations. Table 17 illustrates some of the engineering characteristics for compacted soil materials according to their Unified classification as made from determination of the Atterberg limits and other engineering soil tests. The tests and the classifications are the means by which predictions of behavior and designs of roads, foundations, embankments, waste disposal systems, and other structures can be made for the different soils.

TABLE 17 *Characteristics of soil materials for compacted embankments according to the Unified classification (adapted from Soil Survey Staff, 1971)*

Unified classification	Compressibility	Compaction characteristics	Permeability of compacted soil
GW	Low	Good	High
GP	Low	Good	High
GM	Low	Fair to good	Medium to low
GC	Low to medium	Good to fair	Low
SW	Low	Good	High
SP	Low	Good	High
SM	Low to medium	Fair to good	Medium to low
SC	Low to medium	Good to fair	Low
ML	Medium	Fair to poor	Medium to low
CL	Medium	Fair to good	Low
MH	High	Poor	Low to medium
CH	High	Fair to poor	Low
OL	High	Fair to poor	Low to medium
OH	High	Poor	Low
Pt	Not suitable	Not suitable	Not suitable

SOIL CLASSIFICATION LAB ANALYSES

Laboratory analyses of soils used for soil classification purposes are designed to describe (characterize) specific soils, and to provide as much information as possible about the genesis of each soil. Pedological (soil survey) analyses are generally very laborious and time-consuming, and require expensive equipment for X-ray diffraction, differential thermal analysis, centrifuging, cutting thin sections, determination of ions by flame photometry, and so on. Table 18 is a list of some important analyses used for soil characterization and classification (Soil Survey Staff, 1967). Because these analyses are designed to completely characterize the soils, they are elaborate and comprehensive. Pedological analyses are designed to determine minute differences in effects of leaching and weathering between horizons of the same soil, as well as major physical, chemical, and minerological properties of vastly different soils in contrasting regions.

Particle-size analyses (Table 18) segregate the different size separates of soil horizons by weight by sieving and sedimentation; the sandy, silty, and clayey components of soils influence most of the other properties. The nature of the particles and mineralogy is determined by X-ray diffraction, optical analysis, surface area determinations, and other investigations. Bulk density shows the weight of the soil per unit volume, and reflects difficulty of root penetration and soil permeability to air and water. Water retention shows the capacity of the soil profile to act as a water storage reservoir for plant growth. Linear extensibility is a measure of the shrink–swell character of the soil with wetting and drying. Thin sections help in the identification of specific minerals in the soils, and in the calculation of pore spaces and films of clay and mineral and organic stains on, within, or between the soil aggregates.

Studies of bases (calcium, magnesium) in the soils (Table 18) give an indication of the nutrient status of each horizon, of weathering processes, and of movement of cations and other materials with the leaching waters over hundreds, thousands, or even millions

TABLE 18 *List of some of the most important laboratory analyses used for soil characterization and classification work (adapted from Soil Survey Staff, 1967)*

Particle-size analyses	Bicarbonate
Bulk density	Chloride
Water retention	Sulfate
Linear extensibility	Nitrate
Thin sections (micromorphology)	Calcium
Cation-exchange capacity	Magnesium
Extractable bases	Sodium
Base saturation	Potassium
Sodium saturation	Sulfur
Organic carbon	Phosphorus
Nitrogen	X-ray diffraction
Iron	Differential thermal analysis
Manganese	Optical analysis
Calcium carbonate	Total analysis
Gypsum	Surface area
Aluminum	Saturation extract
Extractable acidity	pH
Carbonate	Ratios of clay, Ca, Mg, etc.

of years. Organic carbon, nitrogen, iron, manganese, and other materials have character-istic distributions in certain soil groups, so that analyses of these materials are vital for classifications of soils. High aluminum is characteristic of some deeply weathered acid soils, and high contents of water-soluble cations and anions (bicarbonate, chloride, sulfate) are typical of soils of arid regions affected by salinity. Other special analyses of soils are also determined in soil characterization laboratories, as special data needs arise in certain soil survey areas. Often knowledge of minute differences in the results of soil analyses can save hundreds of thousands of dollars in fertilizer costs and engineering construction in soil survey areas.

Soil Test Kit

Techniques of soil analyses are not static, but are in a continual state of standardiza-tion and improvement. During the development of the criteria for soil classification in the Soil Taxonomy system (Soil Survey Staff, 1975), for example, soil test kits were de-veloped that could be used in the field to analyze important soil constituents. One of these portable field soil test kits is pictured in Figure 26. This soil test kit uses a minimal amount of chemical reagents through the use of plastic syringes in which tiny volumes of solutions can be precisely measured; the syringes can be easily cleaned in the field with small amounts of demineralized water. The kit shown in Figure 26 has a spring balance

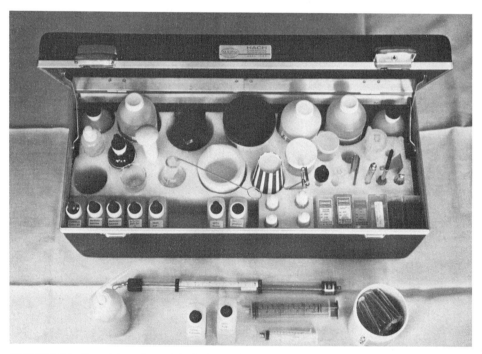

FIGURE 26 *Portable soil test kit designed for use in the field. The spring balance can be read to the nearest 0.01 g with a vernier scale. The test kit enables use of minimal amounts of reagents with plastic syringes, and utilizes exchange resin pillows for reaction measurements. This soil test kit will analyze soil samples for the determinations listed in Table 19.*

that reads to the nearest 0.01 g with a vernier scale. Exchange resin pillows (resin enclosed in plastic tubes) can be easily prepared for chemical exchange reactions, and indicators and buffers make the determinations accurate and simple.

The portable soil test kit (Fig. 26; Hach, 1973) analyzes the extract of a saturated soil paste. Total soluble salts (Table 19) are determined by first titrating the carbonate and bicarbonate, then passing the extract through a hydrogen form resin. Titration of the released acidity gives a value equivalent to the total soluble salts. Chloride is determined on the same solution by another titration. Calcium plus magnesium is measured by a direct titration on a separate aliquot, and sodium is estimated from the difference between total soluble salts and calcium plus magnesium. Sulfate is found as the difference between total soluble salts and all determined anions, or it can be determined directly on a separate aliquot. Several derivative values used in characterizing and classifying saline and sodic soils can be calculated from the analytical values, including sodium absorption ratio, exchangeable sodium percentage, and total soluble salts (ppm and percent of soil). Electrical conductivity of the saturation extract can be estimated from the meq/liter total soluble salts.

The cation-exchange properties of a soil can be determined with the field test kit (Fig. 26) by extraction of a soil sample with a potassium chloride–triethanolamine buffer at pH 8.1. A single extract can be used to determine both buffer extractable acidity and calcium plus magnesium; under most conditions the sum of these two values is a good measure of cation-exchange capacity. In soils of arid regions, high sodium content can be determined by leaching a soil sample with saturated gypsum solution and noting the decrease in calcium content. Neutral salt extractable acidity is determined (Hach, 1973) by neutralizing the buffer before passing it through the soil; the acidity in this extract is caused primarily by exchangeable aluminum. Gypsum is determined by grinding the soil with water to dissolve the gypsum and analyzing the decanted solution; the dissolved

TABLE 19 *Soil test determinations that can be performed on soil samples in the field with the portable soil test kit pictured in Figure 26 (Hach, 1973)*

Qualitative test for soluble salts	Cation-exchange properties
Total salts in 1:2 extract	Buffer extractable acidity
Soluble salts in saturation extract	Calcium + magnesium
Saturation percentage	Cation-exchange capacity
Carbonate	Base saturation
Carbonate + bicarbonate	Calcium
Bicarbonate	Magnesium
Total soluble salts	Sodium (gypsum requirement)
Chloride	Aluminum
Sulfate (estimated)	Gypsum
Electrical conductivity of	Calcium carbonate equivalent
saturation extract (estimated)	
Sulfate	
Calcium + magnesium	
Calcium	
Magnesium	
Sodium (estimated)	
Sodium absorption ratio	
Exchangeable sodium percentage	

gypsum is precipitated with acetone, redissolved in water, and titrated as calcium. Calcium carbonate is determined by measuring the volume of carbon dioxide generated by reaction of a measured weight of soil with a strong acid.

Laboratory analyses of soils are of tremendous value, but the most important data about soils are the soil profile descriptions which characterize the soils in the field in the landscape context where they occur. Laboratory data supplement the soil profile descriptions, and both field and lab data are needed to classify and interpret the soils for use and management. Fertility, engineering, and classification lab analyses are most common— but other analyses can also be used where special problems demand them. Soils are complex entities in our landscapes, and we must have a great deal of information about them if we are to use them efficiently and wisely.

CHAPTER 4

SOIL CLASSIFICATION

All complex natural materials must be classified if they are to be understood. Soil classification is the technique by which soils can be aggregated into categories that are useful for understanding genesis, properties, and behavior. All soils in the United States (more than 11,000 in 1980) and numerous soils in many other countries have been classified according to Soil Taxonomy (Soil Survey Staff, 1975). The Soil Taxonomy classification system was specially designed to be useful at different levels of detail and generalization. Tables 20–25 provide an introduction to the nomenclature of Soil Taxonomy, and Figures 27–30 illustrate different levels of detail and generalization of application of the soil classification system to description of land areas.

Soil Taxonomy is a natural comprehensive system of soil classification, which classifies soil properties for many uses (as completely as is presently feasible). Soils are defined as discrete bodies produced by the interactions of climate, organisms (especially vegetation), and surficial geologic materials in landscapes which are increasingly influenced by human activities. The basic data for the taxonomy consists of soil profile (pedon) descriptions (Olson, 1976) made in the field, and physical and chemical laboratory analyses (Soil Survey Staff, 1967). Soil Taxonomy is a tool designed to serve the soil mapping programs, so that characteristics of soil can be translated into maps showing areal distribution of attributes which influence past, present, and future land use and management.

TABLE 20 *Soil Order names and their formative elements (adapted from Buol et al., 1973, and Soil Survey Staff, 1975)*

Order	Formative element	Derivation	Mnemonicon
Vertisol	ert	L.—*verto*, turn	in*vert*
Entisol	ent	(nonsense syllable)	re*cent*
Inceptisol	ept	L.—*inceptum*, beginning	inception
Aridisol	id	L.—*aridus*, dry	ar*id*
Spodosol	od	Gr.—*spodos*, wood ashes	Po*d*zol; o*dd*
Ultisol	ult	L.—*ultimus*, last	*ult*imate
Mollisol	oll	L.—*mollis*, soft	mo*ll*ify
Alfisol	alf	(nonsense syllable)	Peda*lf*er
Oxisol	ox	Fr.—*oxide*, oxide	*ox*ide
Histosol	ist	Gr.—*histos*, tissue	h*ist*ology

53

TABLE 21 *Formative elements of soil Suborder names (adapted from Buol et al., 1973, and Soil Survey Staff, 1975)*

Formative element	Derivation	Mnemonicon	Meaning or connotation
alb	L.—*albus*, white	*alb*ino	Presence of albic horizon (a bleached eluvial horizon)
and	Modified from *Ando*	*And*o	Andolike
aqu	L.—*aqua*, water	*aqu*arium	Characteristics associated with wetness
ar	L.—*arare*, to plow	*ar*able	Mixed horizons
arg	Modified from argillic horizon; L.—*argilla*, white clay	*arg*illite	Presence of argillic horizon (a horizon with illuvial clay)
bor	Gr.—*boreas*, northern	*bor*eal	Cool
ferr	L.—*ferrum*, iron	*ferr*uginous	Presence of iron
fibr	L.—*fibra*, fiber	*fibr*ous	Least decomposed stage
fluv	L.—*fluvius*, river	*fluv*ial	Floodplains
hem	Gr.—*hemi*, half	*hem*isphere	Intermediate state of decomposition
hum	L.—*humus*, earth	*hum*us	Presence of organic matter
lept	Gr.—*leptos*, thin	*lept*ometer	Thin horizon
ochr	Gr.—base of *ochros*, pale	*ochr*er	Presence of ochric epipedon (a light-colored surface)
orth	Gr.—*orthos*, truc	*orth*ophonic	The common ones
plag	Modified from Ger. *Plaggen*, sod		Presence of plaggen epipedon
psamm	Gr.—*psammos*, sand	*psamm*ite	Sand textures
rend	Modified from Rendzina	*Rendz*ina	Rendzinalike
sapr	Gr.—*sapros*, rotten	*sapr*ophyte	Most decomposed stage
torr	L.—*torridus*, hot, dry	*torr*id	Usually dry
trop	Modified from Gr.— *tropikos*, of the solstice	*trop*ical	Continually warm
ud	L.—*udus*, humid	*ud*ometer	Of humid climates
umbr	L.—*umbra*, shade	*umbr*ella	Presence of umbric epipedon (a dark-colored surface)
ust	L.—*ustus*, burnt	comb*ust*ion	Of dry climates, usually hot in summer
xer	Gr.—*xeros*, dry	*xer*ophyte	Annual dry season

NOMENCLATURE

Soil Taxonomy (Soil Survey Staff, 1975) is a hierarchical system with six categories: Order, Suborder, Great Group, Subgroup, Family, and Series. Each category is designed to be useful for a given purpose at an appropriate level of detail or generalization. Soil Orders, at the highest level, are listed in Table 20. Figure 27 shows the general distribution of Orders in a map of soils of the world. Soil Series, at the lowest level, consist of specific soil names such as the "Glenelg" series in the District of Columbia (Table 25). On detailed soil maps (Fig. 29) the Glenelg series is subdivided further into landscape units such as "Glenelg loam, 0 to 8 percent slopes," "Glenelg loam, 8 to 15 percent slopes," and "Glenelg loam, 15 to 25 percent slopes."

The usefulness of Soil Taxonomy is perhaps best illustrated by example of the nomenclature. The Glenelg series (Table 25; Smith, 1976) consists of deep well-drained soils formed in residuum from micaceous schist and gneiss. The Glenelg series is classified as a "Typic Hapludult; fine-loamy, mixed, mesic." The Subgroup name is "Typic Haplu-

TABLE 22 *Formative elements of soil Great Group names (adapted from Buol et al., 1973, and Soil Survey Staff, 1975)*

Formative element	Derivation	Mnemonicon	Meaning or connotation
acr	Modified from Gr.—*Akros*, at the end	*acro*lith	Extreme weathering
agr	L.—*ager*, field	*agr*iculture	An agric horizon
alb	L.—*albus*, white	*alb*ino	An albic horizon
and	Modified from *Ando*	*And*o	Andolike
anthr	Gr.—*anthropos*, man	*anthr*opology	An anthropic epipedon
aqu	L.—*aqua*, water	*aqu*arium	Characteristic associated with wetness
arg	Modified from argillic horizon; L.—*argilla*, white clay	*arg*illite	An argillic horizon
calc	L.—*calcis*, lime	*calc*ium	A calcic horizon
camb	L.—*cambiare*, to exchange	*c*ha*m*ge	A cambic horizon
chrom	Gr.—*chroma*, color	*chrom*a	High chroma
cry	Gr.—*Kryos*, cold	*cry*stal	Cold
dur	L.—*durus*, hard	*dur*able	A duripan
dystr	Modified from Gr.—*dys*, ill; *dystrophic*, infertile	*dystr*ophic	Low base saturation
eutr	Modified from Gr.—*eu*, good; *eutrophic*, fertile	*eutr*ophic	High base saturation
eu			
ferr	L.—*ferrum*, iron	*ferr*ic	Presence of iron
frag	Modified from L.—*fragillis*, brittle	*frag*ile	Presence of fragipan
fragloss	Compound of *fra(g)* and *gloss*		See the formative elements *frag* and *gloss*
gibbs	Modified from *gibbsite*	*gibbs*ite	Presence of gibbsite
gloss	Gr.—*glossa*, tongue	*gloss*ary	Tongued
hal	Gr.—*hals*, salt	*hal*ophyte	Salty
hapl	Gr.—*haplous*, simple	*hapl*oid	Minimum horizon
hum	L.—*humus*, earth	*hum*us	Presence of humus
hydr	Gr.—*hydro*, water	*hydr*ophobia	Presence of water
hyp	Gr.—*hypnon*, moss	*hyp*num	Presence of hypnum moss
luo, lu	Gr.—*louo*, to wash	ab*lu*tion	Illuvial
moll	L.—*mollis*, soft	*moll*ify	Presence of mollic epipedon
nadur	Compound of *na(tr)* and *dur*		
natr	Modified from *natrium*, sodium		Presence of natric horizon
ochr	Gr.—base of *ochros*, pale	*och*er	Presence of ochric epipedon (a light-colored surface)
pale	Gr.—*paleos*, old	*pale*osol	Old development
pell	Gr.—*pellos*, dusky		Low chroma
plac	Gr.—base of *plax*, flat stone		Presence of a thin pan
plag	Modified from Ger.—*Plaggen*, sod		Presence of plaggen horizon
plinth	Gr.—*plinthos*, brick		Presence of plinthite
quartz	Ger.—*quarz*, quartz	*quartz*	High quartz content
rend	Modified from Rendzina	*Rend*zina	Rendzinalike
rhod	Gr.—base of *rhodon*, rose	*rhod*odendron	Dark red colors
sal	L.—base of *sal*, salt	*sal*ine	Presence of salic horizon
sider	Gr.—*sideros*, iron	*sider*ite	Presence of free iron oxides
sombr	Fr.—*sombre*, dark	*somb*er	A dark horizon
sphagno	Gr.—*sphagnos*, bog	*sphagn*um moss	Presence of sphagnum moss
torr	L.—*torridus*, hot and dry	*torr*id	Usually dry
trop	Modified from Gr.—*tropikos*, of the solstice	*trop*ical	Continually warm
ud	L.—*udus*, humid	*ud*ometer	Of humid climates
umbr	L.—base of *umbra*, shade	*umbr*ella	Presence of umbric epipedon
ust	L.—base of *ustus*, burnt	com*bust*ion	Dry climate, usually hot in summer
verm	L.—base of *vermes*, worm	*verm*iform	Wormy, or mixed by animals
vitr	L.—*vitrum*, glass	*vitr*eous	Presence of glass
xer	Gr.—*xeros*, dry	*xer*ophyte	Annual dry season

dult" (Table 25) and the Order of "Ultisol" is indicated by the formative element "ult" (Table 20) at the end of the Subgroup name. The Suborder of "Udult" is indicated by the last two syllables of the Subgroup name. The Great Group designation is "Hapludult." The Family modifiers are "fine-loamy, mixed, mesic." Thus, by diagram, the Glenelg soil listed in Table 25 from the District of Columbia is classified as (Soil Survey Staff, 1975)

Family modifiers	Subgroup name
Fine-loamy, mixed, mesic	Typic Hapludult

ult...Order, Ultisol

udult...Suborder, Udult

Hapludult...Great Group, Hapludult

The "Ultisol" Order (Fig. 28) indicates that the soil is usually moist and has a horizon of clay accumulation and a low base supply. The "Udult" Suborder indicates that the soil has low organic-matter content in a humid climate (Table 21) with temperate or warm temperatures (Fig. 28). The "Hapludult" Great Group shows that the soil is of the simple type in this category (Table 22). The "typic" part of the Subgroup name indicates that the soil is typical of that group; other formative elements of Subgroup

TABLE 23   *Formative elements of soil Extragrade Subgroups used in the Soil Taxonomy names and their meanings (adapted from Soil Survey Staff, 1975)*

Objective	Meaning
Abruptic	A large difference in percentage of clay between an eluvial horizon and an illuvial horizon without a significant transitional horizon
Aeric	Browner and better aerated than typic
Anthr	A man-made dark-colored surface horizon
Arenic	Sandy eluvial horizons (sand or loamy sand), mostly between 50 cm and 1 m thick
Cumulic	An overthickened epipedon rich in humus
Glossic	Tongued eluvial and illuvial horizons
Grossarenic	Sandy eluvial horizons (sand or loamy sand) more than 1 m thick
Hydric	Organic soil floating on water if used in name of a Histosol
Leptic	Thin soil horizons
Limnic	Organic soil with basal layer of marl, diatoms, or sedimentary peat
Lithic	Hard rock within 50 cm of the surface
Pergelic	Presence of permafrost
Petrocalcic	An indurated horizon of lime accumulation
Petroferric	A shallow layer of ironstone
Pachic	A thick dark surface horizon
Plinthic	Presence of small amounts of plinthite, an iron-rich material that hardens irreversibly on exposure
Ruptic	Intermittent horizons
Sulfic	Presence of deep sulfides or moderate amounts if shallow, or products of sulfide oxidation
Superic	Very shallow plinthite
Terric	A mineral substratum in an organic soil
Thapto	A buried soil

names are listed in Table 23. The Family modifiers of "Fine-loamy, mixed, mesic" designate a specific (Table 24, Fig. 14) particle-size class, mixed mineralogy, and a mean annual soil temperature between 8 and 15°C (mesic). All of these soil characteristics are significant to use and management of land.

The 10 soil Orders (Table 20) can be recognized in a soil Family name from the last syllable of the Subgroup part. The Order is the broadest category of the Soil Taxonomy system (Figs. 27 and 28). The Suborder name has two syllables: the last is the Order formative element, and the first designates additional diagnostic properties of those soils. Great Group names have three or four syllables: the last two are the name of the Suborder, and the first denotes other diagnostic characteristics. Great Groups are separated into three kinds of subgroups: Typic, Intergrade, and Extragrade (Table 23) — a Typic Subgroup represents the central concept of its Great Group, an Intergrade Subgroup has the definite properties of the Great Group whose name it carries as a substantive plus

TABLE 24 *Modifiers that express particle-size classes in soil Family names (adapted from Soil Survey Staff, 1975)*

Class	Definition
Fragmental	Stones, cobbles, gravel, and very coarse sand particles; too little fine earth to fill interstices > 1 mm in diameter
Sandy-skeletal	Rock fragments 2 mm or coarser make up 35% or more by volume; enough fine earth to fill interstices > 1 mm; the fraction < 2 mm is sandy as defined for "Sandy" particle-size class
Loamy-skeletal	Rock fragments make up 35% or more by volume; enough fine earth to fill interstices > 1 mm; the fraction < 2 mm is loamy as defined for "Loamy" particle-size class
Clayey-skeletal	Rock fragments make up 35% or more by volume; enough fine earth to fill interstices > 1 mm; the fraction finer than 2 mm is clayey as defined for "Clayey" particle-size class
Sandy	The texture of the fine earth is sand or loamy sand but not loamy very fine sand or very fine sand; rock fragments make up < 35% by volume
Loamy	The texture of the fine earth is loamy very fine sand, very fine sand, or finer, but the amount of clay (carbonate clay considered as silt) is < 35%; rock fragments are < 35 % by volume
Coarse-loamy	By weight, 15% or more of the particles are fine sand (0.25–0.1 mm) or coarser, including fragments up to 7.5 cm in diameter; < 18% clay in the fine-earth fraction
Fine-loamy	By weight, 15% or more of the particles are fine sand (0.25–0.1 mm) or coarser, including fragments up to 7.5 cm in diameter; 18–34% clay in the fine-earth fraction (< 30% in Vertisols)
Coarse-silty	By weight, < 15% of the particles are fine sand (0.25–0.1 mm) or coarser, including fragments up to 7.5 cm in diameter; < 18% clay in the fine-earth fraction
Fine-silty	By weight, < 15% of the particles are fine sand (0.25–0.1 mm) or coarser, including fragments up to 7.5 cm in diameter; 18–34% clay in the fine-earth fraction (< 30% in Vertisols)
Clayey	The fine earth contains 35% or more clay by weight, and rock fragments are < 35% by volume
Fine-clayey	A clayey particle-size class for soils having 35–59% clay in the fine-earth fraction (30–59% for Vertisols)
Very-fine	A clayey particle-size class for soils having 60% or more clay in the fine-earth fraction

TABLE 25 *Soil Taxonomy classification of the soils of the District of Columbia (adapted from Smith, 1976)*

Soil name, symbols	Family or Great Group taxonomic class
Ashe, As	Coarse-loamy, mixed, mesic Typic Dystrochrept
Beltsville, Bd, Be, Uc	Fine-loamy, mixed, mesic Typic Fragiudult
Bibb, Bg	Coarse-loamy, siliceous, acid, thermic Typic Fluvaquent
Bourne, Bn, Bp	Fine-loamy, mixed, thermic Typic Fragiudult
Brandywine, Br, Bt, Ud	Sandy-skeletal, mixed, mesic Typic Dystrochrept
Chillum, Cc, Cd, Ue	Fine-silty, mixed, mesic Typic Hapludult
Christiana, Ce, Cf, Uf	Clayey, kaolinitic, mesic Typic Paleudult
Codorus, Ck, Cn	Fine-loamy, mixed, mesic Fluvaquentic Dystrochrept
Croom, Cw, Cx, Uk	Loamy-skeletal, mixed, mesic Typic Hapludult
Dunning, Dn	Fine, mixed, mesic Fluvaquentic Haplaquoll
Fallsington, Fa	Fine-loamy, siliceous, mesic Typic Ochraquult
Fluvaquents, FB, FD	Fluvaquents
Galestown, Ge, Um	Sandy, siliceous, mesic Psammentic Hapludult
Glenelg, Gg, Gh	Fine-loamy, mixed, mesic Typic Hapludult
Glenelg Variant, Gl, Gm	Fine-loamy, mixed, mesic Aquic Hapludult
Iuka, Ik, Ip	Coarse-loamy, siliceous, acid, thermic Aquic Udifluvent
Joppa, Jt, Ju, Uo	Loamy-skeletal, siliceous, mesic Typic Hapludult
Keyport, Ke, Km, Up	Clayey, mixed, mesic Aquic Hapludult
Lindside, Ld, Ip	Fine-silty, mixed, mesic Fluvaquentic Eutrochrept
Manor, Mb, Mc, Md, Us	Coarse-loamy, micaceous, mesic Typic Dystrochrept
Metapeake, Mg	Fine-silty, mixed, mesic Typic Hapludult
Melvin, Mp	Fine-silty, mixed, nonacid, mesic Typic Fluvaquent
Muirkirk Variant, Mv	Coarse-loamy over clayey, siliceous, mesic Typic Paleudult
Neshaminy, Ne, Nu	Fine-loamy, mixed, mesic Ultic Hapludalf
Rumford, Gf	Coarse-loamy, siliceous, thermic Typic Hapludult
Sassafrass, Sa, Sc, Sg, Ux	Fine-loamy, siliceous, mesic Typic Hapludult
Sunnyside, Sm, Sp, Uy	Fine-loamy, siliceous, mesic Typic Hapludult
Udorthents, Ul-II	Udorthents
Udifluvents, UA	Udifluvents
Woodstown, We, Wp, Uz	Fine-loamy, siliceous, mesic Aquic Hapludult

some of the properties of another taxon (Order, Suborder, or Great Group), and an Extragrade Subgroup has aberrant properties representing Intergrades to materials not considered to be soil (hard rock, permafrost, etc.). The Families of mineral soils are named with the Subgroup designation plus several modifiers that narrow the range of properties to permit general statements about use and management of the areas; Family modifiers may include several designations, including those for particle size, mineralogy, calcareous status and reaction (pH), soil temperature, soil depth, soil slope, soil consistence, coatings, and class of permanent cracks. Thus, Soil Taxonomy permits a wide range of generalization and detail in the use of its hierarchical classes; all the classes are defined precisely on the basis of soil data (Soil Survey Staff, 1975). The systematic logic of the system is heavily weighted toward soil genesis and the practical application of the soil information to use and management of land areas. Several decades of worldwide development and testing by many soil scientists have culminated in the recent publication of Soil Taxonomy (Soil Survey Staff, 1975); additional changes and refinements will be made in the system in the future as more knowledge is acquired about the various kinds of soils in the world. So far, more than 1,000 years of combined individuals' efforts have been invested in the Soil Taxonomy system.

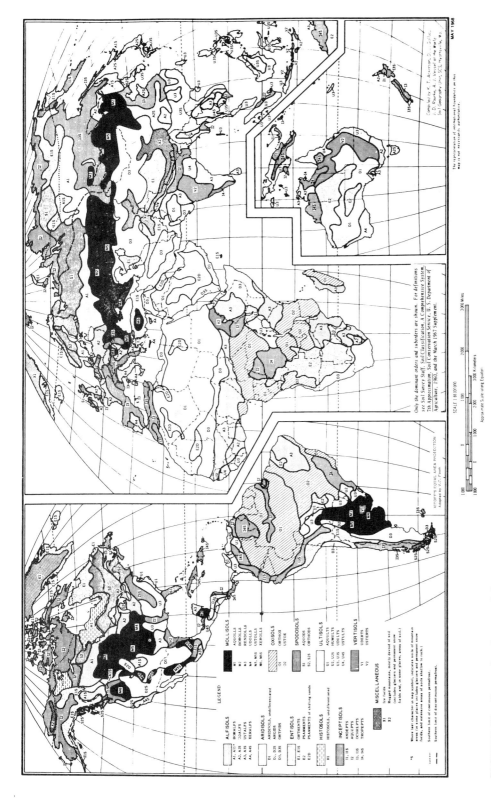

FIGURE 27 *General soil map of the world, showing Orders and Suborders according to Soil Taxonomy (map from Soil Survey Division, Soil Conservation Service, U.S. Department of Agriculture, and Brady, 1974).*

59

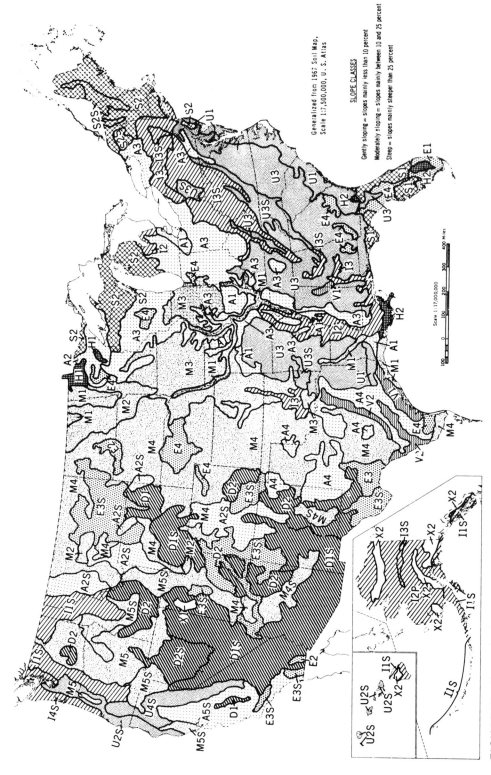

FIGURE 28 General soil map of the United States showing Orders and Suborders according to Soil Taxonomy (map from Soil Survey Division, Soil Conservation Service, U.S. Department of Agriculture, and Brady, 1974).

U. S. DEPARTMENT OF AGRICULTURE

LEGEND

Only the dominant orders and suborders are shown. Each delineation has many inclusions of other kinds of soil. General definitions for the orders and suborders follow. For complete definitions see Soil Survey Staff, Soil Classification, A Comprehensive System, 7th Approximation, Soil Conservation Service, U. S. Department of Agriculture, 1960, (for sale by U. S. Government Printing Office,) and the March 1967 supplement (available from Soil Conservation Service). Approximate equivalents in the modified 1938 soil classification system are indicated for each suborder.

ALFISOLS . . . Soils with gray to brown surface horizons, medium to high base supply, and subsurface horizons of clay accumulation; usually moist but may be dry during warm season

A1   AQUALFS (seasonally saturated with water), ntly sloping; general crops (if drained), pasture and woodland if undrained (Some Low—Humic Gley soils and Planosols)

A2   BORALFS (cool or cold) gently sloping; mostly woodland, pasture, and some small grain (Gray Wooded soils)

A2S   BORALFS steep; mostly woodland

A3   UDALFS (temperate or warm, and moist) gently or moderately sloping; mostly farmed, corn, soybeans, small grain, and pasture (Gray—Brown Podzolic soils)

A4   USTALFS (warm and intermittently dry for long periods) gently or moderately sloping; range, small grain, and irrigated crops (Some Reddish Chestnut and Red—Yellow Podzolic soils)

A5S   XERALFS (warm and continuously dry in summer for long periods, moist in winter) gently sloping to steep; mostly range, small grain, and irrigated crops (Noncalcic Brown soils)

ARIDISOLS . . . Soils with pedogenic horizons, low in organic matter, and dry more than 6 months of the year in all horizons

D1   ARGIDS (with horizon of clay accumulation) gently or moderately sloping; mostly range, some irrigated crops (Some Desert, Reddish Desert, Reddish—Brown, and Brown soils and associated Solonetz soils)

D1S   ARGIDS gently sloping to steep

D2   ORTHIDS (without horizon of clay accumulation) gently or moderately sloping; mostly range and some irrigated crops (Some Desert, Reddish Desert, Sierozem, and Brown soils, and some Calciusols and Solonchak s ' ls)

D2S   ORTHIDS gently sloping to steep

ENTISOLS . . . Soils without pedogenic horizons

E1   AQUENTS (seasonally saturated with water) gently sloping; some grazing

E2   ORTHENTS (loamy or clayey textures) deep to hard rock; gently to moderately sloping; range or irrigated farming (Regosols)

E3   ORTHENTS shallow to hard rock; gently to moderately sloping; mostly range (Lithosols)

E3S   ORTHENTS shallow to hard rock; steep; mostly range

E4   PSAMMENTS (sand or loamy sand textures) gently to moderately sloping, mostly range in dry climates, woodland or cropland in humid climates (Regosols)

HISTOSOLS . . . Organic soils

H1   FIBRISTS (fibrous or woody peats, largely undecomposed) mostly wooded or idle (Peats)

H2   SAPRISTS (decomposed mucks) truck crops if drained, idle if undrained (Mucks)

INCEPTISOLS . . . Soils that are usually moist, with pedogenic horizons of alteration of parent materials but not of accumulation

I1S   ANDEPTS (with amorphous clay or vitric volcanic ash and pumice) gently sloping to steep; mostly woodland, in Hawaii mostly sugar cane, pineapple, and range (Ando soils, some Tundra soils)

I2   AQUEPTS (seasonally saturated with water) gently sloping; if drained, mostly row crops, corn, soybeans, and cotton; if undrained, mostly woodland or pasture (Some Low—Humic Gley soils and Alluvial soils)

I2P   AQUEPTS (with continuous or sporadic permafrost) gently sloping to steep; woodland or idle (Tundra soils)

I3   OCHREPTS (with thin or light-colored surface horizons and little organic matter) gently to moderately sloping, mostly pasture, small grain, and hay (Sols Bruns Acides and some Alluvial soils)

I3S   OCHREPTS gently sloping to steep; woodland, pasture, small grains

I4S   UMBREPTS (with thick dark-colored surface horizons rich in organic matter) moderately sloping to steep; mostly woodland (Some Regosols)

MOLLISOLS . . . Soils with nearly black, organic-rich surface horizons and high base supply

M1   AQUOLLS (seasonally saturated with water) gently sloping; mostly drained and farmed (Humic Gley soils)

M2   BOROLLS (cool or cold) gently or moderately sloping, some steep slopes in Utah; mostly small grain in North Central States, range and woodland in Western States (Some Chernozems)

M3   UDOLLS (temperate or warm, and moist) gently or moderately sloping; mostly corn, soybeans, and small grain (Some Brunizems)

M4   USTOLLS (intermittently dry for long periods during summer) gently to moderately sloping; mostly wheat and range in western part, wheat and corn or sorghum in eastern part, some irrigated crops (Chestnut soils and some Chernozems and Brown soils)

M4S   USTOLLS moderately sloping to steep; mostly range or woodland

M5   XEROLLS (continuously dry in summer for long periods, moist in winter) gently to moderately sloping; mostly wheat, range, and irrigated crops (Some Brunizems, Chestnut, and Brown soils)

M5S   XEROLLS moderately sloping to steep; mostly range

SPODOSOLS . . . Soils with accumulations of amorphous materials in subsurface horizons

S1   AQUODS (seasonally saturated with water) gently sloping; mostly range or woodland, where drained in Florida, citr.s and special crops (Ground-Water Podzols)

S2   ORTHODS (with subsurface accumulations of iron, aluminum, and organic matter) gently to moderately sloping; woodland, pasture, small grains, special crops (Podzols, Brown Podzolic soils)

S2S   ORTHODS steep; mostly woodland

ULTISOLS . . . Soils that are usually moist with horizon of clay accumulation and a low base supply

U1   AQUULTS (seasonally saturated with water) gently sloping; woodland and pasture if undrained, feed and truck crops if drained (Some Low—Humic Gley soils)

U2S   HUMULTS (with high or very high organic—matter content) moderately sloping to steep, woodland and pasture if steep, sugar cane and pineapple in Hawaii, truck and seed crops in Western States (Some Reddish—Brown Lateritic soils)

U3   UDULTS (with low organic—matter content; temperate or warm, and moist) gently to moderately sloping; woodland, pasture, feed crops, tobacco, and cotton (Red—Yellow Podzolic soils, some Reddish—Brown Lateritic soils)

U3S   UDULTS moderately sloping to steep; woodland, pasture

U4S   XERULTS (with low to moderate organic—matter content, continuously dry for long periods in summer) range and woodland (Some Reddish—Brown Lateritic soils)

VERTISOLS . . . Soils with high content of swelling clays and wide deep cracks at some season

V1   UDERTS (cracks open for only short periods, less than 3 months in a year) gently sloping, cotton, corn, pasture, and some rice (Some Grumusols)

V2   USTERTS (cracks open and close twice a year and remain open more than 3 months), general crops, range, and some irrigated crops (Some Grumusols)

AREAS with little soil . . .

X1   Salt flats

X2   Rockland, ice fields

NOMENCLATURE

The nomenclature is systematic. Names of soil orders end in sol (L. solum, soil), e. g. ALFISOL, and contain a formative element used as the final syllable in names of taxa in suborders, great groups, and sub-groups.

Names of suborders consist of two syllables, e. g. AQUALF. Formative elements in the legend for this map and their connotations are as follows:

and   — Modified from Ando soils; soils from vitreous parent materials

aqu   — L. aqua, water; soils that are wet for long periods

arg   — Modified from L. argilla, clay; soils with a horizon of clay accumulation

bor   — Gr. boreas, northern; cool

fibr   — L. fibra, fiber; least decomposed

hum   — L. humus, earth; presence of organic matter

ochr   — Gr. base of ochros, pale; soils with little organic matter

orth   — Gr. orthos, true; the common or typical

psamm   — Gr. psammos, sand; sandy soils

sapr   — Gr. sapros, rotten; most decomposed

ud   — L. udus, humid; of humid climates

umbr   — L. umbra, shade; dark colors reflecting much organic matter

ust   — L. ustus, burnt; of dry climates with summer rains

xer   — Gr. xeros, dry; of dry climates with winter rains

AUGUST 1967

FIGURE 28 (continued)

MAPS

Figure 27 is probably the best single map to use for teaching about the broadest and most general aspects of Soil Taxonomy (Soil Survey Staff, 1975). Figure 28 is a similar type of map, but provides more map detail and definitive descriptions in the legend. Many maps at different scales are available or can be made to broaden or narrow the perspectives of the distributions of soils on landscapes from place to place. Figure 29 is an example of a detailed soil map about which rather precise statements can be made on the classifications of soils and their uses in the landscapes.

In the District of Columbia (Washington, D.C.) a variety of different soils exist which have influenced the development of the city. Table 25 is a list of the soils in the District and the Soil Taxonomy classifications. Figure 29 is a detailed map sheet (made at 1:12,000 scale) for the area around Brightwood, which is about 5 miles north of the White House. The left (west) part of the Map Sheet 3 has soils formed mostly in residuum from saprolite (weathered material) of crystalline rocks of the Piedmont; soils in the right (east) portion of the map sheet formed mostly in sediments of the Coastal Plain. Figure 30 is a schematic diagram of some soils on the "Fall Line" between the Piedmont Plateau and the Coastal Plain. The Fall Line marks a position where terrain changes and is now the boundary of the two geomorphic regions which runs north-south through the middle of Figure 29. The soil map illustrates very well the diversity of the two different soils landscapes.

In Figure 29 the soils in Rock Creek Park are shown to be mostly Manor (Mb) and Glenelg (Gg) on steep slopes (C, 8-15 percent; D, 15-25 percent). Manor is a coarse-loamy, micaceous, mesic Typic Dystrochrept in saprolite from crystalline rocks. The steep slopes retarded urban development, so that the area could be relatively easily preserved as parkland; the area is very scenic and provides a place for biking and other recreational activities. Much of the area on gentler slopes has been urbanized with houses and other buildings. Several soils are mapped as Udorthents of different textures, because the soils have been much disturbed by construction activities. The UxB map unit in the eastern portion of Map Sheet 3 (Fig. 29), for example, has fine-loamy, siliceous, mesic Typic Hapludults formed in loamy coarse sand, but the sandy materials have been disturbed during the construction of the buildings and streets. Figure 30 is a schematic diagram of the Fall Line transition area, including some soils not mapped in Figure 29, but showing the general nature of the substrata of soils in the landscapes adjacent to the Fall Line.

The beauty of Soil Taxonomy (Soil Survey Staff, 1975) is in the elegance of its scientific logic. Soils are important entities of our natural environment, but they are elusive for categorization because of their complexities. Only with such a massive effort over many years could such an advancement be accomplished as is exemplified in Soil Taxonomy. The value of Soil Taxonomy will become apparent when the subject matter and detail of its contents are widely understood and extensively used by everyone.

GROUPINGS OF SOILS

Users cannot comprehend all of the complexities of soils without grouping them into meaningful units. Soil groupings involve methods and criteria by which specific narrowly defined soils and map units at the lowest levels can be placed into broader classes or categories necessary for making management and design decisions about uses of land. Soil

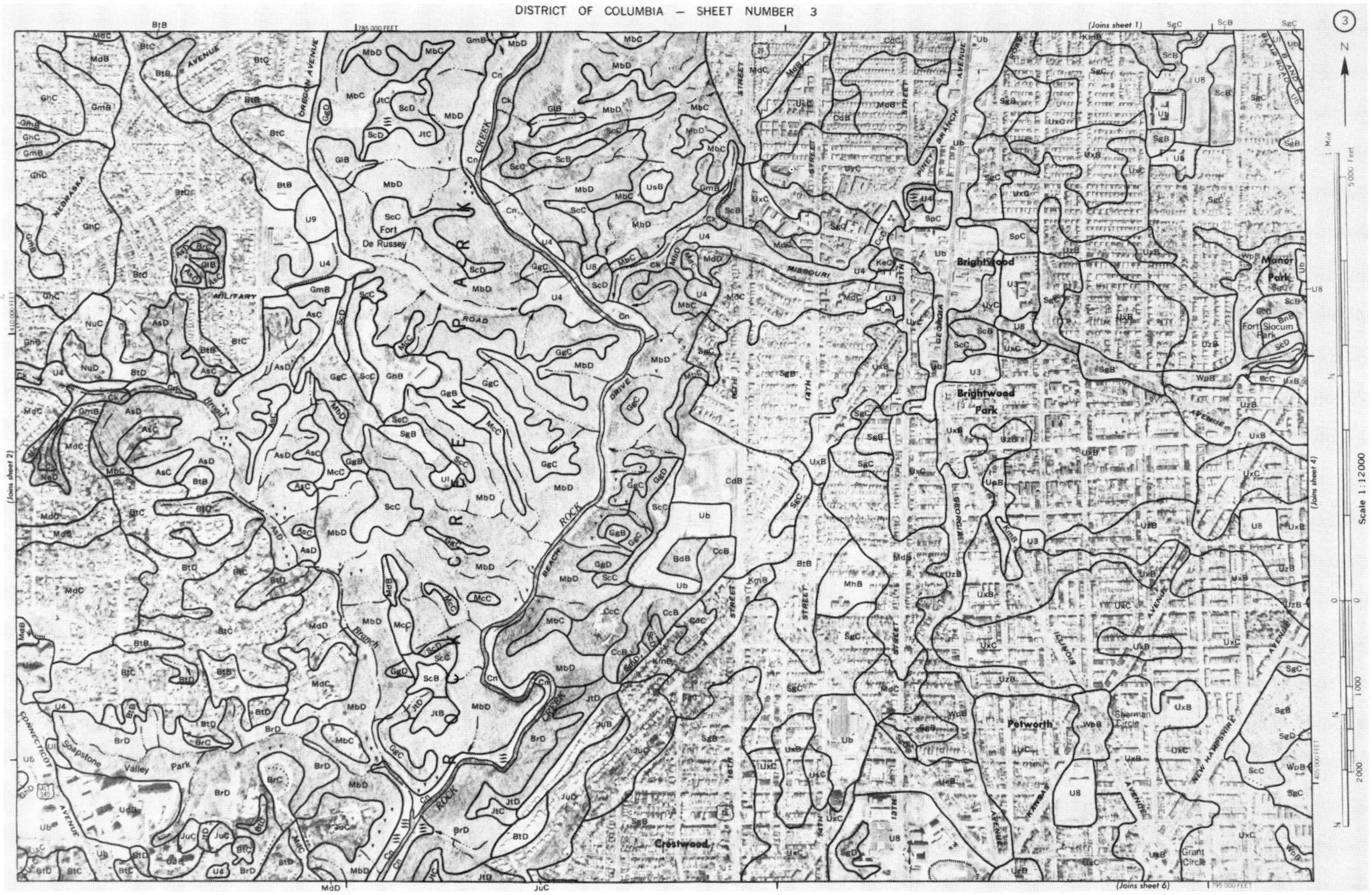

FIGURE 29 Detailed Soil Map Sheet 3 of the area around Brightwood in the District of Columbia, about 5 miles north of the White House (map reduced from scale of 1:12,000; Smith, 1976).

63

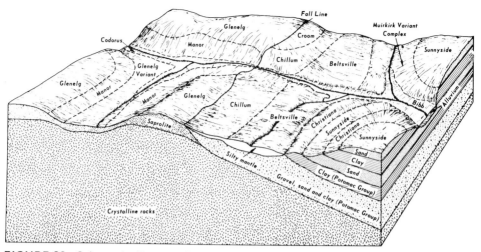

FIGURE 30 *Schematic diagram of geomorphology and soils adjacent to the Fall Line between the Piedmont Plateau and the Coastal Plain in the District of Columbia (adapted from Smith, 1976).*

groupings have been termed "technical groupings" (Orvedal and Edwards, 1941) because the classes are defined by the technical descriptions of the soil properties. Technical groupings are a narrowing of the soil classification concepts and are generally based on only a few important characteristics of soils, and can be readily adapted to detailed soil maps. Soil groupings may be "artificial" or "arbitrary" groupings of soil properties that are important for a specific use. Groupings of soils may change as technology is improved, as when tracked vehicles are introduced into an area to enable tillage before traditional wheeled vehicles could traverse the fields, or when improved herbicides enable tillage to be reduced or eliminated at critical seasons when soils are wet.

Groupings of soils involve "real" and "inferred" soil properties and ratings of soils for specific purposes. "Real" properties are those which can be readily described, measured, and mapped—such as soil color, texture, structure, consistence, and reaction. Inferred soil properties are those that are not observed directly, but can be "inferred" from the characteristics that can be measured. Thus, soil erosion cannot be measured when the soil profile is described, because the topsoil that has been eroded is missing. However, a soil scientist can readily and accurately "infer" that the soil profile has lost topsoil, because he or she knows the characteristics of the nearby uneroded soil profiles and also has observed the sediment deposited downslope. Similarly, permeability of soil to air and water is generally inferred (although it can be measured in the field) from the real soil properties of texture, structure, and so on. Obviously, a wet clayey soil with poor structure can be "inferred" to have slow permeability to air and water, without measuring permeability at very many sites (measured permeability at a few sites can be "extended" to many sites of the same mapped soil). This principle of "data transfer" permits data collected at a few sites to be extended (within reasonable limits) to other areas of the same or similar mapped soils.

Drainage Classes

One grouping of soils that has proved to be extremely useful is that of categorizing soils according to their drainage classes (Soil Survey Staff, 1962). Figure 31 is a diagram showing some of the soil drainage classes. The rooting zone and the colors of the soils can be easily measured, but the depth to water is inferred from soil mottling and colors that indicate oxidized and reduced materials. Soil water table fluctuations may be measured at a few sites, but collecting such data is very time-consuming when numerous measurements are taken over periods of years.

In well-drained soils (Fig. 31), water is readily removed from the soil by runoff or internal drainage; the soil is free of mottles in the upper part. In humid regions these well drained soils are usually moist, but only rarely very wet or very dry. Classes of excessively drained soils are sometimes used where soils are droughty because of shallowness to bedrock, very steep slopes, or coarse textures that limit the moisture-holding capacity for plant growth. In moderately well drained soils, water is removed from the soil somewhat slowly, so that the profile is occasionally wet for short but significant periods. Moderately well drained soils generally have slowly permeable layers in the B or C horizons, a seasonal high water table in the B, and accumulations of seasonal seepage waters. In somewhat poorly drained soils that are wet, water is removed from the soil slowly enough to keep it wet for significant periods but not all of the time; these soils have a slowly permeable layer, a high water table, and accumulate water through seepage. Poorly drained soils are usually wet, and water is removed so slowly that the soil remains wet for much of the year. The water table in poorly drained soils is near the surface during much of the year, a slowly permeable layer is present, and seepage is a problem.

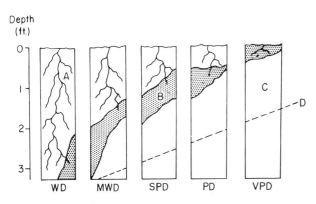

A	Rooting zone
B	Mottled zone
C	Gray, brown, or stained zone
D	Permanent or fluctuating water table

WD	Well drained
MWD	Moderately well drained
SPD	Somewhat poorly drained
PD	Poorly drained
VPD	Very poorly drained

FIGURE 31 *Soil drainage classes and some associated real and inferred soil properties.*

Very poorly drained soils are continuously wet, and the water table remains near the surface most of the year. These soils occupy depressions and ponded sites, where water accumulates. If wet soils are permeable, of course, they can be drained if an outlet is available and much improved for various uses. If wet soils are only slowly permeable, then drainage and amelioration are difficult.

Interpretive Maps

Many interpretative soil maps can be made to show "single-factor" soil conditions of real or inferred soil properties. Maps of groupings of soils, for example, can be made to show drainage class or depth to water table. Similar maps can be made to show texture, stoniness, rockiness, slopes, shallow depth to bedrock, permeability, foundation bearing strength, water-holding capacity, shrink–swell, rooting depth, trafficability, erosion, and other soil factors. Often the "single-factor maps" are especially effective in interpreting soil maps for planning, because the maps can be readily understood. Thus, a map showing areas with bedrock at 20 in. or less, a map showing areas with high water tables at less than 20 in. for most of the year, a map showing steep slopes of 35–45 percent, a map showing areas with soils of pH less than 5.0, and other maps can very effectively show the limitations of development in different places. Overlay maps can be made of different transparencies so that hazards of different kinds can be emphasized and suitable soil areas (without hazards) can be located. With different presentation techniques, the negative aspects (poor soil conditions) can be emphasized, or the positive approach can be followed to outline the good soil areas and the corrective measures needed to make the poor soils better for specific uses.

Soils are generally grouped to show suitability, limitations, or capabilities for specific purposes. Table 26 is one example of groupings of soils for growing plants, based on soil drainage class, moist consistence, texture, slope, thickness, and volume of coarse fragments. For cropping or landscape design, a number of soil properties are important. Categories can be made for each soil characteristic that is important for a specific use. Limits of categories for the classes can be adapted to soils of a specific area, or they can be standardized for a large region or country. Often, the most limiting soil property is

TABLE 26 *Ratings of soils for growing plants (adapted from Soil Survey Staff, 1971)*

Item affecting use	Soil suitability rating		
	Good	Moderate	Poor
Soil drainage class	Well and moderately well drained	Somewhat poorly drained	Poorly and very poorly drained
Moist consistence	Very friable, friable	Loose, firm	Very firm, extremely firm
Textures	vfsl, fsl, l, sil, sl	cl, scl, sicl, sc	s, ls, c, sic
Thickness of soil (above hard layer, water table, or bedrock)	> 30 in.	20–30 in.	< 20 in.
Coarse fragments (volume)	< 8%	8–15%	> 15%
Slope	< 8%	8–15%	> 15%

used to rate the entire soil (soil profile or soil map unit), but some limiting soil characteristics are easier to correct than others. A poorly drained soil, for example, would be rated "poor" in Table 26, but the rating could be "moderate" or "good" if the water management could feasibly be improved. A sandy soil would be "poor" even if well drained, due to the texture because of droughtiness for plant growth, but it could be improved with irrigation. Some soils, of course, are very poorly drained, very firm clays that would be more difficult to improve for plant growth.

It is relatively easy to place soils into three groups: those that are good, moderate, or poor for a specific purpose (or have slight, moderate, or severe limitations). The good soil characteristics are usually readily apparent (Table 26). The poor soil properties are also generally obvious to most observers. When the "good" and "poor" soils are segregated, the rest of the soils without the extreme characteristics can be lumped into a middle or "moderate" class (Table 26). Thus, a three-class system of rating soils is relatively simple and easy to understand and conservative—because the differences are extreme and obvious.

TABLE 27 *Ratings of soils for general farming in New York State (adapted from Rogoff, 1976)*

Item affecting use	Soil Potential				
	Very good	Good	Moderate	Poor	Very poor
Drainage class and approximate depth in inches to permanent or fluctuating water table	Well drained > 36	Moderately well drained 18–36	Somewhat poorly drained 12–18	Poorly drained 6–12	Very poorly drained < 6
Total water-holding capacity (in. H_2O/rooting depth)	> 6	4–6	3–4	2–3	< 2
Slope (%)	< 3	3–8	8–15	15–25	> 25
Rooting depth (in. to root restricting horizon)	> 40	30–40	20–30	10–20	< 10
Trafficability (Unified soil group)	GW, GP, SW, SP, GM, GC, SM, SC, Pt (drained)	CL with PI < 15	ML, CL with PI > 15	OH, OL, MH	CH, Pt (undrained)
Permeability class (in./hr. in least permeable horizon)	0.6–2.0	2.0–6.0	> 6.0	0.6–0.06	< 0.06
Erosion	None	Slightly eroded	Moderately eroded	Severely eroded	Very severely eroded
Stoniness and Rockiness	None	Few	Some	Many	Very many
pH in B horizon	> 7.0	6.5–7.0	6.0–6.5	5.5–6.0	< 5.5
Texture	sil	l, sicl	sl, cl	scl, c	s, ls (not irrigated)
Elevation (ft)	< 400	400–1000	1000–1600	1600–2000	> 2000

For many purposes, simple groupings of soil characteristics into three categories are not adequate. Table 27 is an illustration of refinements that can be made. Table 27 is a rating system like that used in Table 26, but the three classes of Table 26 have been refined into five classes in Table 27—and more soil characteristics have been added to the criteria for constituting the classes. Soil drainage class, depth to water table, water-holding capacity, rooting depth, permeability, and texture are all related to moisture storage for crops, and ease of root penetration into the soil. Slope, stoniness, and rockiness interfere with operation of heavy machinery on the soil map units. The trafficability classes are based on the Unified engineering classification and measure the ease or difficulty with which heavy machinery can traverse soil map units under varying soil moisture conditions. The pH in the B horizon (Table 27) is a measure of the natural fertility of the soil; the pH and nutrients in the A horizon can be improved relatively easily with additions of lime and fertilizers. Elevation of soil map units is related to length of growing season; high elevations may have frost-free growing seasons several weeks shorter and generally cooler temperatures in most years than soils in the lower valleys.

The criteria by which soils are rated for different uses are based on practical experience in describing and mapping the soils, and in observing the land use on the soils in the survey area. Figures 32-34 provide some examples of correlations between soils and land use. In Figure 32, the soils on the levee ridges along the bayou in Terrebonne Parish, Louisiana, are mostly Mhoon (Md, Mf) and Sharkey (Sd). Most of the houses and roads are concentrated on the Mhoon soils, but the pipelines cross both the Mhoon and Sharkey areas. The Mhoon soils are somewhat poorly drained, and the Sharkey soils are poorly drained (Fig. 31). Mhoon soils are stratified alluvial silt loam and silty clay, and the Sharkey soils are mostly clay. Soils farther from the levee are swamp mucky clays (Sg), muck (Sh), peat (Sk), and freshwater marsh with deep peat (Fd). Obviously, the houses and roads are located on the best soils in the area that are highest and best drained in the landscape, the wetter mineral soils are farmed back from the levee, but the very wet organic soils are so poor for development and have so many limitations that they are little used in this environment. Thus, the soil use fits in very well with the soil capability, and the rating system can be easily set up in accord with the experiences of people in using their soils. In other areas of Terrebonne Parish, other soils are moderately well drained and have better characteristics for houses and roads—but each rating system is limited by the universe of the soil characteristics with which it deals.

Both conformity and nonconformity of land uses with soil conditions are illustrated in Figure 33 from an area in Hockley County, Texas. The area has mostly well-drained Amarillo fine sandy loam (AfA) on 0-1 percent slopes, but poorly drained Randall clay (Ra) occupies the playas. Amarillo soils provide a relatively stable engineering base for houses and roads, but Randall clays have a high content of montmorillonite, which causes the soils to swell and shrink with wetting and drying. Some of the Randall (Ra) areas in the city of Levelland have been largely avoided for construction because of the poor soil conditions. Experiences of people with these soils have taught them about the shrink-swell properties and the wet conditions. On the other hand, the roads and railroads follow the grid survey pattern almost without deviation, and cross the Randall (Ra) areas. Many examples of roadbed difficulties could be cited, and eventually the highway and railway engineers learned to excavate and fill the playa areas at great cost. The Amarillo soil is good for farming, but erosion by wind and water pose serious problems. Irrigation has greatly increased the productivity of some of the Amarillo soils during recent years. Many

FIGURE 32 *Soil Map Sheet 17 of the area surrounding Humphreys about 50 miles southwest of New Orleans in Louisiana (adapted from Lytle et al., 1960).*

FIGURE 33 *Soil Map Sheet 39 of the area surrounding Levelland about 100 miles south of Amarillo in Texas (adapted from Grice et al., 1965).*

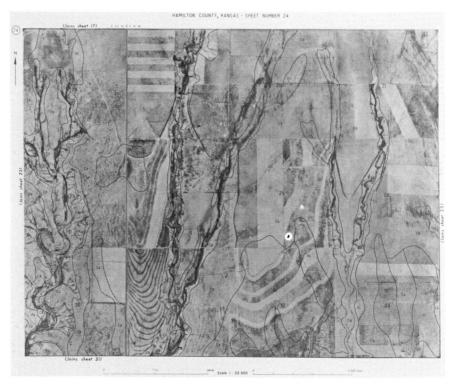

FIGURE 34 *Soil Map Sheet 24 of an area about 240 miles west of Wichita in western Kansas (adapted from McBee et al., 1961).*

problems and limitations of soils can be overcome with expensive inputs. Wet soils can be drained, droughty soils can be irrigated, and montmorillonite clays that shrink and swell can be avoided for construction—or the material can be removed and replaced by fill that does not shrink and swell and provides a better roadbed.

Figure 34 provides an example of how soils can be managed and improved to increase yield potential and conserve the soil resources. The area shown in Hamilton County, Kansas, has mostly Colby silt loam on 1-3 percent slopes. The Colby soil has formed in deep silty loess that is extremely erosive, both from the actions of wind and water. The soil is not well suited for use as cropland, but much of it is being cultivated without irrigation nevertheless. Wheat and sorghum are grown under a system of alternate crop and fallow. The soil tends to seal over on the surface during rainstorms, so that runoff and soil erosion are excessive. When dry, soil particles are easily detached by the strong winds where the soil is not adequately protected by plants or residue. The soil can be conserved, however, by strip-cropping, contour cultivation, and terraces. The conservation measures are well illustrated in Figure 34 on the Colby soils in the northwest quadrant of Soil Map Sheet 24 (strip-cropping), in the southwest (terraces), and in the southeast (strip-cropping). The strip-cropping helps protect the soil from erosion by the prevailing winds, contour cultivation reduces both wind and water erosion, and terraces very effectively prevent excessive runoff water and conserve the soil. The most effective soil conservation measures, of course, are the most expensive to put on the land. Strip-cropping is relatively inexpensive; contour cultivation requires survey equipment, technical assistance, and farm layout adjustments; and terraces require expensive earth-moving equipment, grassed waterways for overflow outlets, maintenance of the terrace structures, and many farming adjustments deviating from the conventional simple block field layouts.

Lagol Area Examples

The utility of soil groupings is illustrated by the series of soil maps in Figures 35-41, using the area around Lagol about 50 miles northwest of Los Angeles as an example. This area (Fig. 35) has a large variety of soils; so many, in fact, that it is difficult to grasp significant differences at a glance. The hills have bedrock variety ranging from soft shale to hard sandstone and also include igneous rocks; thus, the soils formed from those rocks are greatly different. Some of the topography is so rough that it has been termed "badland" (BdG soil map unit) and landslides are common in the soils underlain by unstable sediments. Soils in the valley are mostly on alluvial fans and terraces in sediments washed from a variety of soils in the hills. The valley soils include Metz loamy sand (MeA), Huerhuero very fine sandy loam (HuB), Mocho loam (MoA), Sorrento silty clay loam (SxA), and Salinas clay loam (SaA). Textures are highly contrasting due to the different sources of materials and the varying velocity of the runoff waters that deposited the sediments. Salts and lime in the soils are also variable and influence the land-use and cropping patterns.

Soils in the valley are intensively cropped and irrigated for vegetables, fruits, citrus, avocados, and flowers (Fig. 35). Urban expansion is encroaching on cropland, and also spreading into the hills. Uplands are used mainly for range. Certain soils are well suited to specific crops, as some of the sandy soils are good for raising strawberries. Soils with high shrink-swell characteristics cause problems for foundations. Some of the soils need drainage. During the dry season, fires are common in the hills and canyons, and severe erosion, runoff, and landslides often follow in the rainy season on the denuded slopes. Runoff and erosion in the uplands often produce serious consequences of flooding and sediment damage in the valley. In total, knowledge of soils in this environment (as in all environments) is vital for managing all of the croplands, rangelands, and urban lands.

The map in Figure 36 is a soil map of the Lagol area without the aerial photographic background. Such maps are useful to show the soil patterns clearly, and for hand-coloring to show different soil groupings. These "overrun" soil map sheets are reproduced on the printing press when the soil survey report and maps are published, and are usually kept on file in the local District office of the Soil Conservation Service. Upon request, these maps are available for reproduction and use. The air photo base of the soil map is useful for understanding the topography and land-use patterns, but the "overrun" map permits a better reading of the symbols and boundaries of the soil map units.

Hydrologic soil groups (Fig. 37) have proved to be valuable for evaluating runoff and infiltration rates from different soil areas. Knowledge of runoff rates is vital in predicting flooding, and thousands of dollars can be saved each year when such soil information is put to use. Four soil groups (A, B, C, and D) are commonly mapped to predict the hydrologic runoff potential of soils (Edwards et al., 1970). The hydrologic soil groups are based on soil properties that influence runoff. The runoff potential is calculated or estimated from water intake at the end of a long-term storm that occurs after prior wetting with plenty of opportunity for swelling of a soil not protected by vegetation.

Hydrologic Soil Group A is composed of soils that have high infiltration rates when

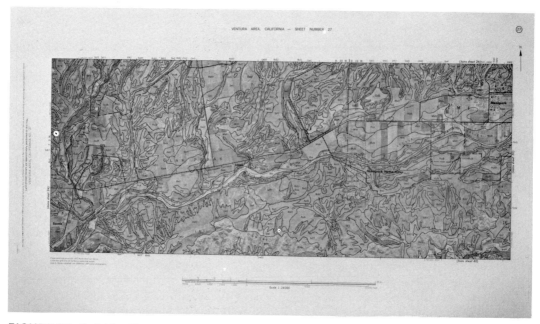

FIGURE 35 *Soil Map Sheet 27 of area surrounding Lagol about 50 miles northwest of Los Angeles, California (adapted from Edwards et al., 1970).*

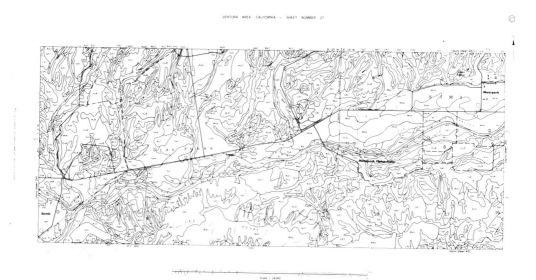

FIGURE 36 *Overrun soil map (without air photo background) made from Soil Map Sheet 27 of area surrounding Lagol, California (adapted from Edwards et al., 1970).*

thoroughly wetted. They are deep, well drained and excessively drained, sand or gravel. Rate of water transmission is high and runoff potential is low.

Hydrologic Soil Group B has soils with moderate infiltration rates when thoroughly wetted. They are deep and moderately deep, well drained and moderately well drained, moderately coarse textured soils. The rate of water transmission through these soils is moderate.

Hydrologic Soil Group C comprises soils with slow infiltration rates when thoroughly wetted. They are soils that have a layer impeding downward movement of water, or moderately fine and fine-textured soils with dense subsoils. Rate of water transmission in the soil profile is slow.

Hydrologic Soil Group D is composed of soils that have very slow infiltration rates when thoroughly wetted. They are clays that have high shrink–swell potential, soils that have a permanent high water table, soils that have a claypan or clay layer at or near the surface, or soils that are shallow over nearly impervious material. Rate of water transmission in these soils is very slow.

Around Lagol (Fig. 37), runoff is very rapid or rapid in the hills, where the infiltration is very slow (Group D) or slow (Group C). Even some soils in the valley on gentle slopes have rapid runoff where the infiltration is slow (Group C). Valley soils in Group B (Fig. 37) have moderate runoff and moderate infiltration. Soils of Group A have high infiltration and low runoff potential, but some of the low-lying areas near the arroyo may be flooded during severe rainstorms (Rw-Riverwash). Such maps can be made for entire survey areas or drainage basins, and are valuable for watershed planning.

Soil erosion potentials (Fig. 38) in the area surrounding Lagol are very high in the steep hills and high in the foothills. Erosion is not much of a problem in the nearly level valley areas, but moderate slopes have some problems. All the factors affecting erosion interact in each environment, so that each area has a unique and distinctive set of soil characteristics and problems. The understanding of runoff and erosion conditions of the soil patterns helps determine the water supply and the human reactions to flood hazards of each community.

The shrink–swell potential of the soils (Fig. 39) is highest in Azule (Au), Chesterton (Ch), and Rincon (Rc) soils on old terraces formed from materials weathered from sedimentary rocks—these soils are relatively high in montmorillonite clays that swell and shrink with wetting and drying. Alluvial soils of younger age generally have low shrink–swell potential. Adjacent soils between the two extremes have moderate shrink–swell potential. Soils with low shrink–swell potential have less than 35 percent clay mainly kaolinite; soils with moderate shrink–swell potential have more than 35 percent kaolinite or somewhat less than 35 percent with some montmorillonite mixed in; and soils with high shrink–swell potential have more than 35 percent clay with much of that likely to be montmorillonite. These considerations (Fig. 39), of course, are critical for behavior of foundations and other engineering structures. Many foundations have been disrupted because people did not realize that the soils had these limitations, or because they did not provide for the proper drainage around their house foundation and other structures. In some places on steep slopes, massive landslides have taken place after soils were saturated and became unstable during heavy rainstorms.

Soil groups for farming (Fig. 40) can be compiled using a variety of techniques. A useful general grouping is derived from the aggregation of the capability classes (Edwards et al., 1970). Very good and good soils for farming in the Lagol area have few or only moderate limitations that deter cropping or require conservation practices; these are deep,

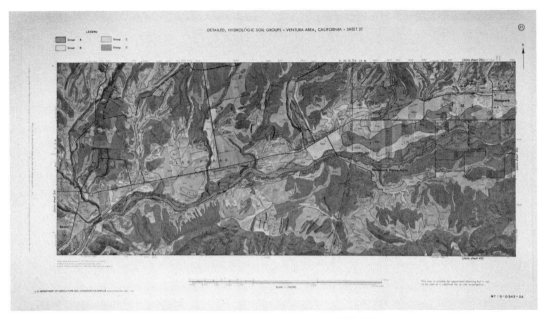

FIGURE 37   *Hydrologic soil groups in areas of Soil Map Sheet 27 of area surrounding Lagol, California (adapted from Edwards et al., 1970).*

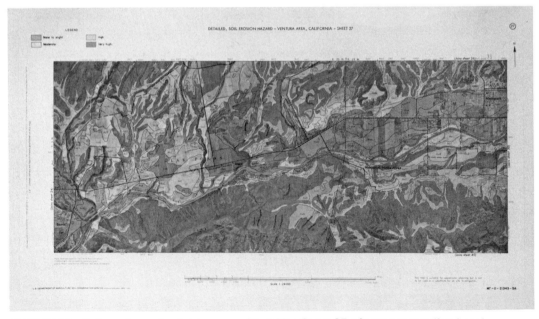

FIGURE 38   *Soil erosion potential in areas of Soil Map Sheet 27 of area surrounding Lagol, California (adapted from Edwards et al., 1970).*

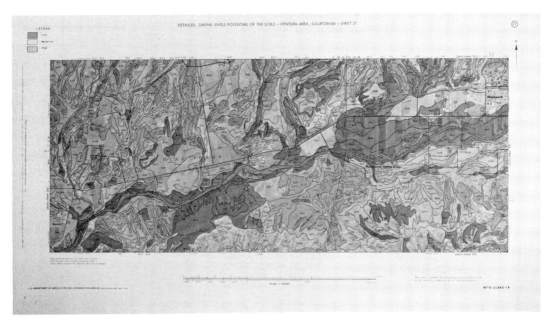

FIGURE 39 *Shrink–swell potential of the soils surrounding Lagol, California (adapted from Edwards et al., 1970).*

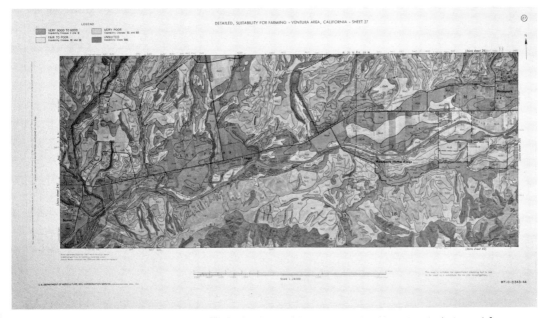

FIGURE 40 *Soil suitability for farming in the area surrounding Lagol, California (adapted from Edwards et al., 1970).*

well-drained soils like Mocho and Sorrento on nearly level slopes. Fair to poor soils for farming include classes III and IV with soils like Azule and Rincon on slopes with severe and very severe erosion potential; these soils need terracing and contouring in places and other expensive management techniques. Very poor soils for farming in capability classes VI and VII are shallow, steep, and stony, and are best left in rangeland. Unsuited areas for farming comprise soils in class VIII, which include the badlands. These groupings of soils include lands so contrasting that present land use conforms rather rigidly to the soil classes. People familiar with the area surrounding Lagol have little argument with the validity of the soil groupings for practical purposes.

Many subtle aspects of soil behavior are not immediately obvious to the casual observer. Avocado root rot (Fig. 41), for example, has a definite correlation between kinds of soil and the fungus *Phytophtora cinnamoni*. The hazards of avocado root rot are directly related to soil drainage and permeability (Edwards et al., 1970). Permeability, in turn, is determined by texture and amount of exchangeable sodium in the subsoils. Soils with slight limitations for avocado root rot are well drained, have rapid or moderate permeability, are deep to hardpan or bedrock, and have a low content of exchangeable sodium. Soils with moderate limitations (Fig. 41) are moderately well drained, permeability is moderately slow in the subsoils, depth to hardpan or impermeable bedrock is between 36 and 60 in., and the content of exchangeable sodium is high. Soils with severe limitations are somewhat poorly drained or poorly drained, have slow subsoil permeability, have hardpan or bedrock at 20 to 36 in., and have a high content of exchangeable sodium. Very severe limitations for avocados are caused by very poor drainage, very slow subsoil permeability, hardpan or bedrock at less than 20 in., and very high exchangeable sodium. Thus, soil characteristics, interpreted into soil groups for practical purposes, can

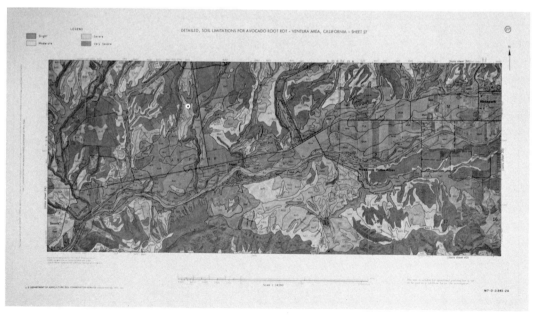

FIGURE 41 *Soil limitations for avocado root rot in the area surrounding Lagol, California (adapted from Edwards et al., 1970).*

save thousands of dollars for developers and managers of avocados and other crops in soil survey areas.

Groupings of soils are techniques to make soil information and maps more understandable to the users of the data. Nobody can immediately comprehend all of the possible soil characteristics given in a lengthy and detailed soil survey report. Soil groups, however, enable planners and developers to consider only those soil properties that are important for a specific use or purpose. Tables can be made to show the reasons for the grouping decisions, and interpretative soil maps show excellently the good, moderate, or poor soil conditions of each soil map unit. These techniques of soil groupings have greatly contributed to making soil information understandable to laypersons, and have also helped soil scientists to better understand the significance of their work. New imaginative processes of computer manipulations of data promise to make soil groupings even more useful in the future.

CHAPTER 5

COMPUTERIZED
DATA PROCESSING

Computers are of tremendous assistance in handling huge amounts of data. The computerization of soil survey data is proceeding at a rapid rate. Most of the data on soils is being manipulated by computers for calculations, classification, correlation, and interpretation—and the computerization trends will continue at an accelerated rate in the future. Most of the details of explanation of the computer systems are in the process of being written for the National Soils Handbook of the Cooperative Soil Survey. The National Soils Handbook is in a state of continual progress and revision, and is available through the local District office and the State office of the Soil Conservation Service, as well as the office of each soil survey representative of the Agricultural Experiment Stations and other contributors to the Cooperative Soil Survey. It is a very lengthy and detailed mimeographed technical manual, and only a small part of the content will be discussed in this chapter.

The most basic data of the Cooperative Soil Survey are the technical soil series descriptions by which each soil is defined and mapped; more than 11,000 soil series were defined and mapped by 1980. Table 28 is an example of the official description for the Norfolk series formed in loamy Coastal Plains sediments from North Carolina to Texas. The Norfolk series is classified as a fine-loamy, siliceous, thermic Typic Paleudult according to Soil Taxonomy (Soil Survey Staff, 1975). The "type location" is in Robeson County, North Carolina, and the range in characteristics is defined for the permissible deviations from the typical. Competing series are given for the geographic setting on the Coastal Plain, and associated soils are listed for those landscapes. Norfolk soils are well drained, and the permeability is moderate. Land use includes cropping for corn, cotton, peanuts, tobacco, and soybeans and original vegetation was pines and mixed hardwoods. The series was first established and mapped in Cecil County, Maryland, in 1900, but many refinements have been made in the description of the soil since 1900. Under "Remarks" is a brief discussion of the history of the classification and definition (description) of the soil. Additional data about the soil includes some references with results of chemical analyses, descriptions, and other information. Format for encoding the technical descriptions for computers are given in the National Soils Handbook. The State of Montana has published a separate booklet on the automated data-processing system for soil inventories in use in that state (Decker et al., 1975). The computer system for soil descriptions use "mark sense forms," which enable cards to be marked with graphite pencils in the field, and then read by electronic scanner machines for recording the soil profile description data into the computer systems.

TABLE 28　*Official soil series description for the Norfolk series*

The Norfolk series consists of deep, well drained, moderately permeable soils that formed in loamy
　Coastal Plains sediments. These upland soils have slopes ranging from 0 to 10%.

Taxonomic Class:　Fine-loamy, siliceous, thermic Typic Paleudults.

Typical Pedon:　Norfolk loam sand — cultivated.
　　　　　　　　　(Colors are for moist soil unless otherwise stated.)

　Ap — 0 to 9 in.; grayish brown (10YR 5/2) loamy sand; weak fine and medium granular structure;
very friable; few fine and medium roots; some darker-colored material in old root channels; strongly
acid; clear smooth boundary. (3 to 10 in. thick)

　A2 — 9 to 14 in.; light yellowish brown (10YR 6/4) loamy sand; weak medium granular structure;
very friable; few fine and medium roots; some darker-colored material in old root channels; strongly
acid; clear smooth boundary. (3 to 10 in. thick)

　B1 — 14 to 17 in.; yellowish brown (10YR 5/6) sandy loam; weak medium subangular blocky
structure; friable; few fine and medium roots; strongly acid; clear wavy boundary. (2 to 5 in. thick)

　B21t — 17 to 38 in.; yellowish brown (10 YR 5/6) sandy clay loam; weak medium subangular
blocky structure; friable; thin discontinuous clay films on faces of peds; many fine and medium pores;
strongly acid; gradual wavy boundary. (12 to 24 in. thick)

　B22t — 38 to 58 in.; yellowish brown (10YR 5/6) sandy clay loam; few fine faint mottles of strong
brown, pale brown, and yellowish red; weak medium subangular blocky structure; friable; thin
discontinuous clay films on faces of peds; strongly acid; gradual wavy boundary. (18 to 24 in. thick)

　B23t — 58 to 70 in.; yellowish brown (10YR 5/6) sandy clay loam; common medium distinct
mottles of yellowish red (5YR 5/8), pale brown (10YR 6/3), and light brownish gray (10YR 6/2);
weak medium subangular blocky structure; friable; few firm yellowish red plinthite nodules; strongly
acid; gradual wavy boundary. (10 to 18 in. thick)

　B3 — 70 to 82 in.; mottled brownish yellow (10YR 6/6), strong brown (7.5YR 5/6), yellowish red
(5YR 5/6) sandy clay loam; weak medium subangular blocky structure; friable; approximately 5%
firm, brittle nodules of plinthite; strongly acid; gradual wavy boundary. (9 to 15 in. thick)

　C — 82 to 100 in.; mottled red (2.5YR 4/8), strong brown (7.5YR 5/8), brownish yellow (10YR
6/8), and gray (10YR 5/1) sandy clay loam; massive; friable; strongly acid.

Type Location:　Robeson County, North Carolina; 1¼ miles south of Parkton; 300 ft. west of State
　Road 1724 and 60 ft. south of farm road.

Range in Characteristics:　The loamy textured horizons commonly extend 60 to 90 in. below the soil
　surface. Few to about 5% small rounded siliceous pebbles are on the surface and throughout the
　soil in some pedons. A few fine or medium rounded ironstone pebbles are present in some pedons.
　Reaction is strongly acid or very strongly acid, except where limed. Mottles, associated with
　seasonal wetness, range from about 36 to 60 in. below the surface.

The A1 horizon ranges from gray to dark grayish brown in hues of 10YR or 2.5Y.

The A2 horizon is very pale brown (10YR 7/3, 7/4), pale brown (10YR 6/3), light yellowish brown
　(10YR 6/4; 2.5Y 6/4), or yellowish brown (10YR 5/4).

The Ap horizon ranges from grayish brown (10YR 5/2) in uneroded pedons to yellowish brown
　(10YR 5/4) or light yellowish brown (10YR 6/4) in eroded pedons. The A horizon is centered on
　loamy sand, and includes fine sandy loam and sandy loam.

The B1 horizon is light yellowish brown (10YR 5/4; 2.5Y 6/4), or yellowish brown (10YR 5/4, 5/6,
　5/8) sandy loam, or sandy clay loam.

The B2t horizon is commonly brownish yellow (10YR 6/6, 6/8), or yellowish brown (10YR 5/6,
　5/8), and ranges to strong brown (7.5YR 5/6, 5/8) or light olive brown (2.5Y 5/4, 5/6). The B2t
　horizon centers on sandy clay loam, and includes sandy loam and clay loam.

The B3 horizon is mottled brownish yellow, strong brown, yellowish red, red, and gray sandy loam,
　sandy clay loam, clay loam or clay. This horizon in some pedons contains firm, brittle strong
　brown to red peds or nodules of plinthite, but no horizon within 60 in. of the soil surface has as
　much as 5% plinthite.

The C horizon is commonly mottled red, strong brown, brownish yellow, and gray loamy soil
　materials, that is variable and may include sand or clay.

Competing Series:　These are the Addielou, Allen, Avilla, Bama, Etowah, Holston, Leesburg, Minvale,
　Nella, Orangeburg, Pikeville, Ruston, and Smithdale series. Addielou soils have A horizons thicker
　than 20 in. Allen, Bama, Etowah, Nella, Orangeburg, Ruston, and Smithdale soils have all or some

TABLE 28   *(continued)*

part of the Bt horizon in hues of 5YR or redder. Avilla, Holston, Leesburg, Minvale, and Pikeville soils have coarse fragments that exceed 10% in all or some part of the solum.

Geographic Setting: Norfolk soils are on nearly level to gently sloping uplands of the Coastal Plain. Slopes range from 0 to 10%. The soil formed in medium to moderately fine textured Coastal Plain sediments. Near the type location, the mean annual temperature is 62° F, and the mean annual precipitation is approximately 49 in.

Geographically Associated Soils: In addition to the competing Orangeburg series, these include the Lakeland, Lynchburg, Rains, and Troup series. Lakeland soils are sandy to depths greater than 72 in. and lack Bt horizons. Lynchburg and Rains soils are more poorly drained and are dominantly gray in the B horizons. Troup soils have sandy A horizons more than 40 in. thick.

Drainage and Permeability: Well drained; slow to medium runoff; moderate permeability. These soils have a seasonally perched water table at depths of 4.0 to 6.0 ft.

Use and Vegetation: Mostly cleared and used for general farm crops, including corn, cotton, peanuts, tobacco, and soybeans. Originally, forests were pines and mixed hardwoods.

Distribution and Extent: Coastal Plain areas of Alabama, Arkansas, Florida, Georgia, Louisiana, Mississippi, North Carolina, South Carolina, Texas, and Virginia. The series is of large extent.

Series Established: Cecil County, Maryland; 1900.

Remarks: The Norfolk series was formerly classified in the Red-Yellow Podzolic great soil group. The series has been greatly restricted in concept since being established, and is no longer recognized in Cecil County, Maryland. The current revision allows a seasonally perched water table at 3.5 to 6 ft. Daniels et al., in a mimeographed publication "Water Table Levels in Some North Carolina Soils," suggest that chroma 3 mottles indicate a fluctuating water table in soils like Norfolk. This soil is typified as having pale brown (10YR 6/3) mottles within 38 in. of the surface.

Additional Data: (1) U.S. Department of Agriculture, "Soil Survey Laboratory Data and Descriptions for Some Soils of Georgia, North and South Carolina." Soil Survey Investigations Report No. 16; SCS, in cooperation with Georgia, North Carolina, and South Carolina Agricultural Experiment Stations; pages 65, 67, 69. (2) U.S. Department of Agriculture, "Certain Properties of Selected Southeastern United States Soils and Mineralogical Procedures for Their Study," Southern Cooperative Series Bulletin 61 (S–14); Soil Conservation Service, Agricultural Research Service and cooperating Experiment Stations; tables 64, 67, 68, (3) U.S. Department of Agriculture, "Selected Coastal Plain Soil Properties," Southern Cooperative Service and cooperating Experiment Stations; pages 40, 42, 44, 46.

SCS FORM 5

Probably the most progress has been made in computerization of soil survey interpretations, and in use of those data in publishing soil surveys and making interpretive maps of soil groupings (Bartelli, 1978; Nichols and Bartelli, 1974). The entire process of publication of soil surveys, in fact, has been greatly accelerated through the use of computers. The form reproduced in Table 29 is the "SCS Form 5," which has been widely used to provide data input on soil survey interpretations into the computer. Table 30 is an example of some of the abbreviated instructions for the computer information to be put on the input form of Table 29. Much of the information on soil description and performance in great detail is in the National Soils Handbook or in SCS "Soils Memoranda," which are part of the standard operating procedures of the Soil Conservation Service and the Cooperative Soil Survey.

The information recorded on the SCS Form 5 includes (Tables 29 and 30) general information on Major Land Resource Area (MLRA), soil map unit, classification, and author of the computer input data. A brief description of the soil is given in a single paragraph. Data on estimated soil properties are listed which are used as the basis for

TABLE 29 *Computer input form for the SCS Form 5 of soil survey interpretations*

TABLE 29 (*continued*)

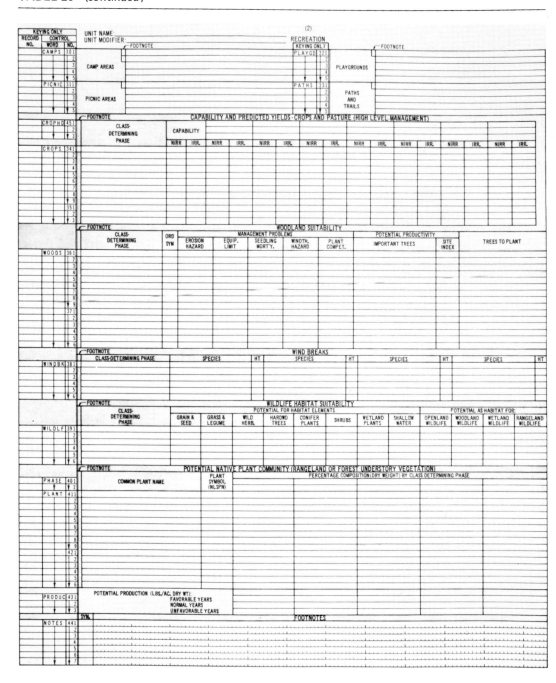

making interpretations of limitations of the soil for various uses. Texture, percent clay, bulk density, permeability, water-holding capacity, reaction, salinity, shrink-swell, erosion factors, organic matter, and corrosivity are listed along with engineering analyses, including percentages of materials passing through sieves of different sizes, liquid limit, plasticity index, and classification of soil horizon materials in the Unified and AASHTO systems (American Association of State Highway and Testing Officials). Information is recorded about frequency and duration of flooding, depths to high water table, cemented pan, bedrock, subsidence, hydrologic group, and potential frost action.

Computer entries (Tables 29 and 30) for sanitary facilities (waste disposal) and community development are based on categories as illustrated in Table 26, where soils are rated as having slight, moderate, or severe limitations for septic tank absorption fields, sewage lagoons, sanitary landfill (trench and area types), and landfill cover. Soil ratings for community development include shallow excavations, dwellings with and without basements, small commercial buildings, and local roads and streets. The slight, moderate, or severe rating is assigned based on the most limiting soil characteristic, and the major restrictive features are stated in the computer language (CEMENTED PAN, SLOPE, WET, etc.). Slight limitations indicate that the soil is relatively free of problems for that use, or limitations can be easily overcome. Moderate limitations need to be recognized, but can be overcome with good management and careful design. Severe limitations are serious enough to make a soil questionable for a specific use. Of course, any soil can be made suitable for any purpose if cost is no object, but economics and feasibility are prime determinants for use of most soil areas.

Use of soils (Tables 29 and 30) as source materials and for water management refer mainly to engineering considerations. Soil areas can be excavated for roadfill, sand, gravel, or topsoil—and soils are rated good, fair, poor, or unsuited for those uses. Obviously, a clay soil would be unsuitable as a source of gravel, and wetness of soils would interfere with excavation for roadfill or topsoil. Limitations of soils are stated for pond reservoir areas and embankments and pond excavations—with limiting soil characteristics listed as DEPTH TO ROCK, UNSTABLE FILL, LARGE STONES, and so on. Ratings of soils and soil map units for drainage, irrigation, terraces, diversions, and grassed waterways generally list only the restrictive soil features, because drainage, irrigation, and waterways often involve watershed planning and water management of other areas in addition to those soil map units being computerized. Thus, the soil features affecting the water management and land use are the most important characteristics for water control. The important soil features affecting water management are listed in the formal computer language in Table 30.

Recreation uses of soils (Tables 29 and 30) are rated according to limitations for camp areas, picnic areas, playgrounds, and paths and trails (Montgomery and Edminster, 1966). Camping areas have severe limitations if wet, flooded, slowly permeable, steeply sloping, or if too sandy or clayey—or if stones and rocks are too numerous. Picnic areas for intensive use have similar soil use considerations except that permeability is not so important; flooding is not as great a hazard for picnic areas as for camping areas. Slopes are major considerations for playgrounds, and paths and trails should not be dusty or wet. Rating tables published in the book on "Soil Surveys and Land Use Planning" by the American Society of Agronomy outline the criteria for soil characteristics for recreational ratings.

Yields (Tables 29 and 30) are predicted for each soil at a high level of management for both irrigated and nonirrigated crops. Capability classes are listed for the different soil

TABLE 30 *Brief instructions for filling out the SCS Form 5 for input of soil interpretation information into the computer system*

SOIL SURVEY INTERPRETATIONS INSTRUCTIONS
(Print clearly in CAPITAL letters. See instructions at bottom of page 2 for entering data legibly.)

1. MLRA(S)--List all the MLRA's (separated by commas) to which the full set of interpretations apply. If some interpretations do not apply to all MLRA's in which the soil occurs, prepare additional SCS-Soils-5's.
2. Kind of Unit--Enter the kind of unit: SERIES, VARIANT, LAND TYPE, GREAT GROUP, SUBGROUP, FAMILY or FAMILY PHASE. If great group, subgroup, family or family phase is entered, the full classification name should be printed on the first line of the block "Classification and Brief Soil Description".
3. Unit Name--Print name of series, variant, land type, etc. (Use no more than 29 characters including spaces). For CNI, if the soil name on diagnostic is in code, print the code in "Unit Name" space.
4. State--Print full name of the state having responsibility for the series.
5. Record No.--State assign a unique sequential number to each Soils-5 prepared, (1 for the first Soils-5 prepared, 2 for the second, etc.) This number is used by the computer to identify each Soils-5. If a Soils-5 is revised, the original record number must be retained.
6. Author(s), Date, Revised--Enter author(s) initials and the date on which the Soils-5 was first prepared or revised. If revised, enter an X in the space provided.
7. Unit Modifier--Enter the soil property which is used as phase criterion that so drastically changes any interpretation that a separate Soils-5 must be prepared, e.g., STONY, SALINE or GRAVELLY SUBSTRATUM.
8. Classification and Brief Soil Description--On the first line (Class 021), print the full classification name of a variant, great group, subgroup, family or family phase. Do not enter classification of series or land types (the computer will print the classification of the series from the current classification file). On lines DESCR 031-035, either print or type (with no more than 120 characters per line and 600 characters total) a short narrative description in non-technical language. This description should not conflict with the standard series description. (1) In the lead sentence give one or two of the major features that characterize the soil and are important to their use and behavior, such as, depth and drainage class. (2) Describe the setting for the series including position on the landscape, shape and ranges of slope names and the parent material where known with reasonable confidence. (3) Describe layers abstracted from the "Estimated Soil Properties" block. Color, texture and thickness of the layers are commonly covered along with special features if they are important to use and management. (4) Give the full range of slope of the series in percent. (See p. 19-20, "Guide to Authors of Manuscripts for Published Soil Surveys".)

ESTIMATED SOIL PROPERTIES
(See "Guide for Interpreting Engineering Uses of Soil" for explanation of many of the following items.)

1. Footnote--Enter a capital letter, e.g., A, if a footnote applies to the whole "Estimated Soil Properties" block. See instructions on page 2 under "footnotes".
2. Depth--A maximum of six layers can be accomodated and up to three sets of surface texture that differ in estimated properties.
3. USDA Texture--Up to three textures can be entered on each line. Separate them by commas. If modifiers are used, they must be attached to the texture by a hyphen, e.g., GR-SL. If a layer is stratified, enter SR as a modifier and the end members of the textural range all connected by hyphens, e.g., SR-S-L.

Modifier:		Texture or terms used in lieu of texture:			
BY	Bouldery	COS	Coarse sand	CE	Coprogenous earth
BYV	Very bouldery	S	Sand	DE	Diatomaceous earth
BYX	Extremely bouldery	FS	Fine sand	FB	Fibric material
CB	Cobbly	VFS	Very fine sand	HM	Hemic material
CBA	Angular cobbly	LCOS	Loamy coarse sand	ICE	Ice or frozen soil
CBV	Very cobbly	LS	Loamy sand	IND	Indurated
CN	Channery	LFS	Loamy fine sand	MARL	Marl
CR	Cherty	LVFS	Loamy very fine sand	MPT	Mucky-peat
CRC	Coarse cherty	COSL	Coarse sandy loam	MUCK	Muck
CRV	Very cherty	SL	Sandy loam	PEAT	Peat
FL	Flaggy	FSL	Fine sandy loam	SP	Sapric material
GR	Gravelly	VFSL	Very fine sandy loam	UWB	Unweathered bedrock
GRC	Coarse gravelly	L	Loam	VAR	Variable
GRF	Fine gravelly	SIL	Silt loam	WB	Weathered bedrock
GRV	Very gravelly	SI	Silt		
MK	Mucky	SCL	Sandy clay loam		
PT	Peaty	CL	Clay loam		
SH	Shaly	SICL	Silty clay loam		
SHV	Very shaly	SC	Sandy clay		
SR	Stratified	SIC	Silty clay		
ST	Stony	C	Clay		
STV	Very stony				
STX	Extremely stony				
SY	Slaty				

4. Unified--Enter up to 4 classes. Separate by commas, e.g., CL-ML, ML. Enter no dual classes except CL-ML.
5. AASHO--Enter up to 4 classes. Separate by commas, e.g., A-6-7, A-7.
6. Fraction>3 in. (Pct)--Enter the weight percentage of material greater than 3 inches, e.g., 30-60. Enter a zero (0) if none occurs.
7. Percent of Material Less Than 3 inches Passing Sieve--Enter the range in weight percentages passing each of the sieve sizes, e.g., 80-100.
8. Liquid Limit--Enter the range of liquid limit, e.g., 20-30, or NP if the soil is nonplastic.
9. Plasticity Index--Enter the range of plasticity index, e.g., 10-20, or NP if the soil is nonplastic.
10. Permeability--Use the following classes: <0.06, 0.06-0.2, 0.2-0.6, 0.6-2.0, 2.0-6.0, 6.0-20, >20. Classes may be combined, e.g., 0.06-0.6.
11. Available Water Capacity--Enter the estimated range of available water capacity in inches per inch, e.g., .10-.15.
12. Soil Reaction (pH)--Enter the range of pH (1:1 water). Commonly the class ranges are given or combination of classes. The classes are 3.5, 3.6-4.4, 4.5-5.0, 5.1-5.5, 5.6-6.0, 6.1-6.5, 6.6-7.3, 7.4-7.8, 7.9-8.4, 8.5-9.0, 9.0.
13. Salinity (mmhos/cm)--Give a range of the electrical conductivity of the saturation extract during the growing season, e.g., 2-10. If salinity is no problem for growing plants, enter a dash.
14. Shrink-Swell Potential--Use one of the following classes: LOW, MODERATE, or HIGH.
15. Corrosivity--Classes are: LOW, MODERATE or HIGH.
16. Erosion Factors--List coordinated K and T factors. List K factors for each major horizon if they are significantly different; T factors for surface layer(s) only. On soils with less than 2 or 3 percent slope, enter a dash.
17. Wind Erodibility Group--Enter wind erodibility group for surface layer(s) only. In parts of the country where wind erosion is no problem, enter a dash.
18. Flooding--Define the natural, unprotected soil in terms of frequency. Duration and months that floods are likely to occur are given only for soils that flood more frequently than rare. Ranges of frequency and duration classes may be given if needed, e.g., RARE-COMMON, BRIEF-V. LONG.

    Frequency: NONE (No reasonable possibility of flooding)
    RARE (Flooding unlikely but possible under abnormal conditions)
    COMMON (Flooding likely under normal conditions)
    OCCASIONAL (Less often than once in 2 years)
    FREQUENT (More often than once in 2 years)

    Duration: V. BRIEF (Less than 2 days)
    BRIEF (2 days to 7 days)
    LONG (7 days to 1 month)
    V. LONG (More than 1 month)

    Months: Give months of probable flooding, use abbreviations, e.g., NOV-MAR.

19. High Water Table--Give the depth range of seasonally high water table to the nearest half-foot, e.g., 1.5-3. Enter APPARENT under kind of water table unless known that the water table is perched, then enter PERCHED. Enter the months in which water table is likely to be within the normal depth of observation, e.g., NOV-MAR. If the water table is below 6 feet or if water table exists for less than one month, enter >6.
20. Cemented Pan--Enter depth range in inches to a cemented pan such as a duripan, petrocalcic, petrogypsic, or ortstein layer. Enter RIPPABLE or HARD in the hardness column. (Rippable and hard are defined under "bedrock" below). If soil has no cemented pan, enter a dash.
21. Bedrock--Enter depth range in inches to bedrock and hardness of rock. Enter RIPPABLE or HARD in the hardness column. "Rippable" rock can be excavated using a single tooth ripping attachment mounted on a 200-300 horsepower tractor. "Hard" rock requires blasting or use of excavators larger than 200-300 horsepower (see Section 2 of Specialification 21, National Engineering Handbook for additional equipment requirments). If depth to bedrock is below the normal depth of observation enter >60.
22. Subsidence--Give depth range in inches of initial and total drainage induced subsidence of organic soils or other wet soils that subside when drained. If subsidence is not a problem, enter a dash.
23. Hydrologic Group--Give the coordinated hydrologic group letter (A,B,C,D,A/D,B/D, or C/D.
24. Potential Frost Action--Enter one of the following: LOW, MODERATE or HIGH. In parts of the country where frost action is no problem, enter a dash.

INTERPRETATIONS FOR SELECTED USES

Give soil limitation or suitability ratings for the selected uses. (Refer to "Guide for Interpreting Engineering Uses of Soil".) Items affecting use that are used as phase criterion are entered in the block first, followed by the limitation or suitability ratings and then the restrictions that apply, e.g., 2-7%: MODERATE-SLOPE. Items used as phase criteria are slope, flooding, texture of surface layer, stoniness and depth to rock. For sloping soils, enter the slope breaks that cause different ratings for each use, e.g., for septic tank absorption fields, the slope breaks are 0-8, 8-15, 15+, for sewage lagoons the slope breaks are 0-2, 2-7, 7+. If flood prone soils are given adequate protection for a given use, rate both the "flooded phase" and the "protected phase", e.g., for dwellings--COMMON: SEVERE-FLOODS and PROTECTED: SLIGHT. Punctuation is important--use colon between phase designation and rating; hyphen between limitations and restrictions; commas to separate any series of items and use no periods. Enter slope first if it is used with additional items affecting use, e.g., 8-15% SCL, CL: MODERATE-SLOPE, TOO CLAYEY. In the following blocks are the rating terms and restrictive features to be used for each use.

SANITARY FACILITIES

Use	Rating	Restrictive Features			
Septic Tank	SLIGHT	CEMENTED PAN	LARGE STONES	SLOPE	
Absorption	MODERATE	DEPTH TO ROCK	PERCS SLOWLY	WET	
Fields	SEVERE	FLOODS	ROCK OUTCROPS		
	SLIGHT	CEMENTED PAN	FLOODS	SLOPE	
Sewage	MODERATE	DEPTH TO ROCK	LARGE STONES	SMALL STONES	
Lagoons	SEVERE	EXCESS HUMUS	PERCS SLOWLY	WET	
Sanitary	SLIGHT	CEMENTED PAN	FLOODS	ROCK OUTCROPS	TOO SANDY
Landfill	MODERATE	DEPTH TO ROCK	LARGE STONES	SLOPE	WET
(Trench)	SEVERE	EXCESS HUMUS	PERCS SLOWLY	TOO CLAYEY	
Sanitary	SLIGHT	FLOODS	WET		
Landfill	MODERATE	PERCS RAPIDLY			
(Area)	SEVERE	SLOPE			
Daily	GOOD	EXCESS HUMUS	PERCS RAPIDLY	THIN LAYER	WET
Cover for	FAIR	HARD TO PACK	SLOPE	TOO CLAYEY	
Landfill	POOR	LARGE STONES	SMALL STONES	TOO SANDY	

COMMUNITY DEVELOPMENT

Use	Rating	Restrictive Features			
Shallow	SLIGHT	CEMENTED PAN	EXCESS HUMUS	ROCK OUTCROPS	TOO CLAYEY
Excavations	MODERATE	CUTBANKS CAVE	FLOODS	SLOPE	WET
	SEVERE	DEPTH TO ROCK	LARGE STONES	SMALL STONES	
Dwellings	SLIGHT	CEMENTED PAN	FLOODS	LOW STRENGTH	SLOPE
Without	MODERATE	DEPTH TO ROCK	FROST ACTION	ROCK OUTCROPS	WET
Basements	SEVERE	EXCESS HUMUS	LARGE STONES	SHRINK-SWELL	
Dwellings	SLIGHT	CEMENTED PAN	FLOODS	LOW STRENGTH	SLOPE
With	MODERATE	DEPTH TO ROCK	FROST ACTION	ROCK OUTCROPS	WET
Basements	SEVERE	EXCESS HUMUS	LARGE STONES	SHRINK-SWELL	
Small	SLIGHT	CORROSIVE	FLOODS	LOW STRENGTH	SLOPE
Commercial	MODERATE	DEPTH TO ROCK	FROST ACTION	ROCK OUTCROPS	WET
Buildings	SEVERE	EXCESS HUMUS	LARGE STONES	SHRINK-SWELL	
Local	SLIGHT	CEMENTED PAN	FLOODS	LOW STRENGTH	SLOPE
Roads and	MODERATE	DEPTH TO ROCK	FROST ACTION	ROCK OUTCROPS	WET
Streets	SEVERE	EXCESS HUMUS	LARGE STONES	SHRINK-SWELL	

REGIONAL INTERPRETATIONS

Interpretations approved for use within the region may be added in these blocks.

SOURCE MATERIALS

Use	Rating	Restrictive Features			
	GOOD	AREA RECLAIM	LARGE STONES	SHRINK-SWELL	WET
Roadfill	FAIR	EXCESS HUMUS	LOW STRENGTH	SLOPE	
	POOR	FROST ACTION	ROCK OUTCROPS	THIN LAYER	
	GOOD				
Sand	FAIR				
	POOR				
	UNSUITED				
	GOOD				
Gravel	FAIR				
	POOR				
	UNSUITED				
	GOOD	AREA RECLAIM	EXCESS SALT	SMALL STONES	TOO SANDY
Topsoil	FAIR	EXCESS ALKALI	LARGE STONES	THIN LAYER	WET
	POOR	EXCESS LIME	SLOPE	TOO CLAYEY	

WATER MANAGEMENT

Use	Rating 1/	Features Affecting			
Pond	SLIGHT 1/	CEMENTED PAN	PERCS RAPIDLY		
Reservoir	MODERATE 1/	DEPTH TO ROCK	SLOPE		
Area	SEVERE 1/	FAVORABLE			
	SLIGHT 1/	COMPRESSIBLE	LARGE STONES	PIPING	UNSTABLE FILL
Pond	MODERATE 1/	FAVORABLE	LOW STRENGTH	ROCK OUTCROPS	
Embankment	SEVERE 1/	HARD TO PACK	PERCS RAPIDLY	THIN LAYER	
Excavated	SLIGHT 1/	DEEP TO WATER	NO WATER		
Ponds	MODERATE 1/	FAVORABLE	ROCK OUTCROPS		
Aquifer Fed	SEVERE 1/	LARGE STONES	SLOW REFILL		
		CEMENTED PAN	DEPTH TO ROCK	FAVORABLE	PERCS SLOWLY
Drainage		COMPLEX SLOPE	EXCESS ALKALI	FLOODS	POOR OUTLETS
		CUTBANKS CAVE	EXCESS SALT	NOT NEEDED	SLOPE
					WET
		COMPLEX SLOPE	EXCESS ALKALI	FAST INTAKE	ROOTING DEPTH
Irrigation		DROUGHTY	EXCESS LIME	FAVORABLE	SLOW INTAKE
		ERODES EASILY	EXCESS SALT	PERCS RAPIDLY	WET
Terraces		CEMENTED PAN	ERODES EASILY	NOT NEEDED	ROCK OUTCROPS
And		COMPLEX SLOPE	FAVORABLE	PERCS SLOWLY	ROOTING DEPTH
Diversions		DEPTH TO ROCK	LARGE STONES	PIPING	SLOPE
				POOR OUTLETS	WET
		DROUGHTY	EXCESS SALT	NOT NEEDED	ROCK OUTCROPS
Grassed		ERODES EASILY	FAVORABLE	PERCS SLOWLY	SLOPE
Waterway		EXCESS ALKALI	LARGE STONES	ROOTING DEPTH	WET

1/ Limitation ratings and restrictions may be used instead of "features affecting" where regional criteria have been developed.

TABLE 30 (continued)

SOIL SURVEY INTERPRETATIONS INSTRUCTIONS

INTERPRETATIONS FOR SELECTED USES (CONTINUED)

RECREATION

Enter the data following the instructions under "Interpretations for Selected Uses". Use Soils Memorandum-69 as a guide in making the ratings. In the following blocks are the rating terms and restrictions features to be used for each use.

Use	Rating		Restrictive Features			Use	Rating		Restrictive Features		
Camp Areas	SLIGHT	DUSTY	PERCS SLOWLY	SMALL STONES	WET	Play Grounds And Trails	SLIGHT	DEPTH TO ROCK	LARGE STONES	SLOPE	TOO SANDY
	MODERATE	FLOODS	ROCK OUTCROPS	TOO CLAYEY			MODERATE	DUSTY	PERCS SLOWLY	SMALL STONES	WET
	SEVERE	LARGE STONES	SLOPE	TOO SANDY			SEVERE	FLOODS	ROCK OUTCROPS	TOO CLAYEY	
Picnic Areas	SLIGHT	DUSTY	ROCK OUTCROPS	TOO CLAYEY		Paths And Trails	SLIGHT	DUSTY	ROCK OUTCROPS	TOO CLAYEY	
	MODERATE	FLOODS	SLOPE	TOO SANDY			MODERATE	FLOODS	SLOPE	TOO SANDY	
	SEVERE	LARGE STONES	SMALL STONES	WET			SEVERE	LARGE STONES	SMALL STONES	WET	

CAPABILITY AND PREDICTED YIELDS - CROPS AND PASTURE (HIGH LEVEL MANAGEMENT)

1. Class-Determining Phase--For each phase that significantly influences yield or management, give the coordinated capability classification and predicted yields of major cultivated crops, hay and pasture commonly grown on the soil. Phases that commonly influence yield and management significantly are flooding, drainage, slope, texture of surface layer and erosion. Enter OCCASIONAL or FREQUENT if flooding influences yield or management. If flood prone soils are given adequate protection for crops, enter PROTECTED in the class-determining phase column and give capability class and subclass and yields. Give capability class and subclass and yields for DRAINED soil if drainage is feasible and UNDRAINED soil if drainage is not feasible. If both drained and undrained phases occur, capability and yields should be given for each. On sloping soils, use the slope groups that are most common in the MLRA in which the typifying pedon is located. Enter SEV.ER. for severely eroded phases that influence yields and management (texture of surface is likely to take on the character of the sub-soil). If moderately eroded phases influence yields and management, enter ERODED. If more than one phase is given, list slope first, texture second, erosion third, e.g., 5-8% CL, SEV.ER. If all phases are rated alike, write ALL in this column.

2. Capability--Give the nonirrigated capability class and subclass for all soils in the column headed NIRR. If the soil is irrigated or is potential irrigated land, enter the irrigated capability class and subclass in the IRR column. Use arabic numbers, 3W, not IIIW.

3. Predicted Yields--On lines CROPHD 451, and 452 if needed, enter the name of the crop. On line CROPHD 453 enter the unit of measure, e.g., (BU), (TONS), (CWT), etc. For each class-determining phase, give the predicted yield of crops approximating those obtained by leading commercial farmers at the level of management which tends to produce the highest economic returns per acre (commonly known as "B" level management). "B" level management includes using the best varieties; balancing plant populations and added plant nutrients to the potential of the soil; control of erosion, weeds, insects and diseases; maintenance of optimum soil tilth; adequate soil drainage; and timely operations. List the yields of the common crops grown in the area in the NIRR (nonirrigated) column, the IRR (irrigated) column, or both columns. Examples of crops and units of measure that are commonly used are as follows:

BARLEY	(BU)	LEGUME HAY	(TONS)	SOYBEANS	(BU)		
CORN	(BU)	OATS	(BU)	SUGAR BEETS	(TONS)		
CORN SILAGE	(TONS)	PASTURE	(AUM)	SUGAR CANE	(TONS)		
COTTON	(LBS LINT)	PEANUTS	(LBS)	TOBACCO	(LBS)		
GRAIN SORGHUM	(BU)	POTATOES	(CWT)	TOMATOES	(TONS)		
GRASS HAY	(TONS)	RICE	(BU)	WHEAT	(BU)		
GRASS-LEGUME HAY	(TONS)	SORGHUM SILAGE	(TONS)				

WOODLAND SUITABILITY

1. If woodland is not an important segment of land use, enter NONE on the first line in the column "Important Trees".
2. Rate only those phases that determine the potential productivity or that are class determining in terms of management. If all phases of a series are rated alike write ALL in the column headed "Class-determining phase".
3. Give only the first two elements of the ordination symbol--the class and the subclass. The third element (group) is regional or local in nature and cannot be coordinated nationally at this time.
4. Follow Soils Memorandum-26, Rev. 2 for guidance in making the ratings, SLIGHT, MODERATE, or SEVERE in the management problem columns.
5. List several of the indicator tree species or forest types and the site index of each in the column "Important Trees". If the site index for a tree species is a summary of 5 or more actual measurements on this soil, enter an asterisk (*) after the index number. A footnote will automatically be printed by the computer as follows: "*Site index is a summary of 5 or more measurements on this soil." Do not enter this footnote on form Soils-5.
6. List one or more tree species suitable for planting. In areas where important, Christmas tree species may be shown in the column "Trees to Plant" followed by a double asterisk, e.g., Arizona Cypress**. A footnote will automatically be printed by the computer as follows: "**Christmas tree species." Do not enter this footnote on form Soils-5. If additional footnotes are needed in this column, they may be entered following the tree species, e.g., SLASH PINE 5/.

WINDBREAKS

Enter the important windbreak tree species and expected height at 20 years for the class-determining phases. If all phases grow the same species, write ALL in the "Class-Determining phase" column. Soils Memorandum-64 contains information on obtaining data for shelterbelts and windbreaks. In parts of the country where windbreaks are not normally needed, enter NONE on the first line under "Species".

WILDLIFE HABITAT SUITABILITY

1. Rate only those phases that are classes determining for potential habitat elements. If all phases of a series are rated alike, write ALL in the "Class-Determining Phase" column. Rating terms are GOOD, FAIR, POOR, V. POOR (to stand for very poor). If an element is not rated, enter a dash.
2. Give a summary rating for the kinds of wildlife. Soils rated for woodland wildlife generally will not be rated for rangeland wildlife and vice versa. If a kind of wildlife is not rated, enter a dash.
3. Use Soils Memorandum-74 as a guide in making ratings.

POTENTIAL NATIVE PLANT COMMUNITY (RANGELAND OR FOREST UNDERSTORY VEGETATION)

1. Enter the names of the class-determining phases for each percentage composition column on line PHASEH 401 and for each percentage composition column on line PHASEH 402, e.g., LOAMY SAND in the first percentage composition column and SANDY LOAM in the second.
2. List the common name of the major native plants that grow under climax condition on this soil and show the plant symbol from the National List of Scientific Plant Names, USDA, SCS, 1971. In rangeland, the species will be grasses, shrubs and forbs and in Savannah sites, trees also are included. Indicate by footnote those rangeland plants that are not usually utilized by cattle or sheep, e.g., YUCCA 6/. In woodland, understory species are listed for grasses, shrubs, forbs and other understory plants within reach of livestock or grazing or browsing wildlife. Understory species and composition should be based on the canopy density which most nearly represents the highest wood production for the forest plant community.
3. For each of the class-determining phases show the percent composition (dry weight) for the major species. Enter OTHER in the plant symbol column and give percentage composition of minor species so that the total is 100 percent.
4. Enter total potential production for favorable years, normal years and unfavorable years.
5. Where data are not available and acceptable estimates cannot be made, list the species in order of their general productivity and leave columns for percent composition blank.
6. Refer to the National Handbook for Range and Related Grazing Lands for additional instructions.

FOOTNOTES

1. Footnote symbols are entered in the "footnote" columns provided in each interpretation block. Enter a capital letter for footnotes that refer to major headings such as Estimated Soil Properties, Sanitary Facilities, etc., and numerals for subordinate headings such as percolation fields, sewage lagoons, etc. The corresponding footnotes should be printed out in the space at the bottom of the second page.
2. No more than one line may be used for any one footnote. For example, to footnote the major heading "Sanitary Facilities" enter a letter A in the footnote column (SEPTIC 071) and on line NOTES 441, enter A in the SYM (symbol) column and print the footnote, RATINGS BASED ON "GUIDE FOR INTERPRETATIONS ENGINEERING USES OF SOILS," NOV. 1971. Footnote the "Septic Tank Absorption Field" heading by entering a number 1 in the footnote column (line SEPT 072), entering a number 1 in the SYM column of line NOTES 442 and print the footnote, EXCESSIVE PERMEABILITY MAY CAUSE POLLUTION OF GROUND WATER. The following are examples of some of the more common footnotes:

ESTIMATES BASED ON ENGINEERING TEST DATA OF (number) PEDONS FROM (counties, states).
RATINGS BASED ON "GUIDE FOR INTERPRETING ENGINEERING USES OF SOILS", NOV. 1971.
EXCESSIVE PERMEABILITY RATE MAY CAUSE POLLUTION OF GROUND WATER.
RECREATION RATINGS BASES ON SOILS MEMORANDUM-69, OCT. 1968.
WILDLIFE RATING BASED ON SOILS MEMORANDUM-74, JAN. 1972.
NOT USUALLY UTILIZED BY CATTLE OR SHEEP.

INSTRUCTIONS FOR ENTERING DATA LEGIBLY FOR KEYPUNCHING EFFICIENCY

1. Make no entries where data are not available or reasonable estimates cannot be made.
2. Do not enter data in "Keying Only" columns.
3. Make entries in black pencil with hardness rating of 2½, H or HB. Entries may be typed, however, take care not to overfill the space (This form has 10 characters per inch, elite type has 12).
4. Print clearly in CAPITAL letters. The following pairs of letters are often confused: AH, LC, UV. Certain letters and numbers are also confused: OO, I1, Z2, S5. Where possibility of confusion exists, use the following symbols for the letters: Ø, I, Z. Make sure the S has round curves and the 5 has sharp corners.
5. Punctuation is important. Follow the instructions and examples given for each of the entries. When using hyphens and decimal points make sure each can be identified.
6. If a block has more than one line, always make entry on first line, use each in order. DO NOT SKIP LINES.
7. If a line data entry in a column is the same below, an arrow can be used to indicate the entry is duplicated, e.g.,

SHRINK-SWELL POTENTIAL

LOW
↓

8. Be sure to PROOF READ.

map units of the same soil. The yields are estimated from all available yield data, in the units listed in Table 30. Each soil map unit has different yield potentials for each crop, and these computer data have proved to be extremely valuable for farm valuation, tax assessment, planning, crop management, and many other purposes.

Woodland suitability (Tables 29 and 30) is used to evaluate soils according to ordination classes and suitability groups. Woodland suitability classes are based on the average site index of an indicator tree species for each soil, with class 1 designating those soils highest in productivity. Site index is the height to which a tree will grow in a given period (e.g., 50 years); often site index curves are used in the yield correlations. The ordination system has three levels: class, subclass, and suitability group. The ordination class is a number that denotes potential productivity in tens of cubic meters of wood per hectare per year for an indicator tree species. The ordination subclass is a letter or letters indicating major soil characteristics or limitations (clayey, restricted rooting, steep, sandy, wet, toxic, stony, rocky). The woodland suitability group (indicated by an Arabic number) is a grouping of soils that have similar potential productivity, grow the same kinds of trees, and have similar limitations or problems in use.

Erosion potential (Tables 29 and 30) for woodlands is a rating for soil erosion following cutting operations where the soil is exposed along roads, skid trails, fire lanes, and log decking areas. Equipment limitations refer to trafficability where soil and slope factors restrict use of tree-harvesting equipment. Seedling mortality is the percentage of seedling mortality (slight 0–25%, moderate 25–50%, severe 50–100%) influenced by soil, erosion, or other site factors. Windthrow hazard evaluates soil characteristics that control tree root development and affect wind firmness of trees. Plant competition is a rating that refers to the invasion or growth of undesirable species when openings are made in the tree canopy. Trees to plant is a listing of species suitable for open field and woodland interplanting.

Trees in windbreaks (Tables 29 and 30) are artificially spaced in rows, and soils under windbreaks have often been altered from their original condition by trapped loess deposits and from effects of the leaf litter remaining on A horizons. Heights of tree species are recorded or estimated after 20 years. Observations of older windbreaks give a good indication about the longevity of trees. Many of the species used in windbreaks have been introduced (Russian olive, Amur honeysuckle, Siberian elm, Nanking cherry) so that their long-term performance on many soils is not well known.

Wildlife habitats (Tables 29 and 30) are rated according to the suitability of the soils for the factors listed in Table 29. Grain and seed crops include annuals such as corn, sorghum, wheat, barley, oats, millet, buckwheat, and sunflower. Domestic grasses and legumes include alfalfa, trefoil, clover, bluegrass, switchgrass, fescue, bromegrass, timothy, orchardgrass, and reed canarygrass. Wild herbaceous upland plants are perennial grasses and weeds that establish themselves naturally and include bluestem, quackgrass, panicgrass, goldenrod, wild carrots, nightshade, and dandelion. Hardwood trees and plants are nonconiferous trees, shrubs, and woody vines that produce nuts or other fruits, buds, catkins, twigs, or foliage that wildlife eat. Hardwoods are also valuable as cover and include native oak, buck, cherry, maple, birch, poplar, apple, hawthorne, dogwood, viburnam, grape, and briars. Planted hardwoods include autumn olive, amur honeysuckle, tatarian honeysuckle, crabapple, multiflora rose, highbush cranberry, and silky dogwood.

Conifer plants (Table 29) are cone-bearing evergreen trees and shrubs that are used mainly for cover by wildlife, including Norway spruce, white pine, white cedar, and hemlock. Soils rated as good for conifer wildlife habitat have shallow depth, very poor drainage, or excessive drainage (droughtiness), which causes conifers to grow very slowly. Slow growth provides dense lower limbs which give good ground cover and protection for wildlife. Good soil character producing slow growth for wildlife habitat is the opposite of good soil character for woodlands, where fast growth is desired.

Wetland plants (Table 29) for wildlife grow on moist-to-wet soils and include smartweed, wild millet, common rush, spikerush, sedges, rice, cutgrass, mannagrass, and cattails. Shallow-water soils are well suited to impoundments with low dikes; they are nearly level, are more than 6 ft to bedrock, are slowly permeable, are poorly to very poorly drained, and have surface runoff or springs as a source of water.

Soils are also rated for wildlife habitat suitability based on general considerations for various kinds of wildlife in openland, woodland, wetland, and rangeland. Some soils are not suited for growing trees, for example, but may be well suited for rangeland wildlife. Other soils may be permanently wetlands which cannot feasibly be drained, and are good for uses by wetland wildlife. Soil ratings for openland wildlife are based on the ratings listed for grain and seed crops, grasses and legumes, wild herbaceous upland plants, hardwood trees and shrubs, and conifer plants. Soil rating for woodland wildlife is based on all the ratings (Table 29) except for grain and seed crops. Rating of soils for wetland wildlife is based on ratings shown for wetland food and cover plants, marshes, and shallow excavated impoundments. Rangeland wildlife ratings are based on the productivity of the native plant range community (Table 29). Openland wildlife includes pheasants, meadowlarks, field sparrows, doves, cottontail rabbits, red foxes, and woodchucks. Woodland wildlife has ruffed grouse, woodcocks, thrushes, vireos, scarlet tanagers, gray and red squirrels, gray foxes, white-tail deer, raccoons, and wild turkeys. Wetlands have ducks, geese, rails, herons, shore birds, minks, muskrats, and beaver. Rangeland wildlife includes antelope, coyote, hawks, and other animals and birds adapted to the native plant communities of the western United States.

Soil interpretations are also made for potential native plant communities (Table 29). Shiflet (1972) summarized procedures for collection, storage, and retrieval of rangeland data by computer systems, and those systems are now in use by the Cooperative Soil Survey in the United States. Annual production is measured and estimated for native plant communities, and the long-term potential production is listed for each soil. Production of rangeland sites is discussed in soil survey reports, according to the soils or soil map units comprising the rangeland sites.

The computer output of the SCS Form 5 soil interpretation data is illustrated in Table 31 for the Norfolk series. The Norfolk soil is a well-drained upland soil in the Coastal Plain that has a loamy profile, with clay accumulations in the B horizons (Table 28). The computer output (Table 31) gives a brief description and data on the soil. Although the soil is well drained, seasonal water causes moderate problems for sanitary facilities due to wetness and seepage in or above the more clayey horizons. The soil is generally of moderate suitability for building-site development on moderate slopes, but some seasonal wetness is an occasional problem. The Norfolk soil is good for roadfill (Table 31), but not good as a source of coarse sand or gravel because of the excess fine sand and loamy and clayey material in the profile. The soil is only moderate for reservoir areas, and has severe limitations for excavated ponds because the seepage is only seasonal.

The Norfolk soil (Table 31) can be easily adapted for recreational uses except where slopes are steep. It is fairly productive for cotton, tobacco, corn, soybeans, wheat, peanuts,and pasture, but needs some liming, fertilization, and erosion control. Norfolk areas are good for growing slash pine and loblolly pine. Norfolk soils are good for openland and woodland wildlife, but very poor for wetland wildlife. Native plant commu-

TABLE 31 *Computer output form for the SCS Form 5 of soil survey interpretations*

NC0037   S O I L   I N T E R P R E T A T I O N S   R E C O R D

MLRA(S): 133, 153, 149   NORFOLK SERIES
REH.WEH. 9-77
TYPIC PALEUDULTS, FINE-LOAMY, SILICEOUS, THERMIC

THE NORFOLK SERIES CONSISTS OF WELL DRAINED NEARLY LEVEL TO SLOPING SOILS ON UPLANDS IN THE COASTAL PLAIN.
IN A REPRESENTATIVE PROFILE, THE SURFACE LAYER IS GRAYISH BROWN LOAMY SAND ABOUT 9 INCHES THICK.
THE SUBSURFACE LAYER IS LIGHT YELLOWISH BROWN LOAMY SAND, 5 INCHES THICK. THE SUBSOIL EXTENDS TO A DEPTH OF 82
INCHES. IT IS YELLOWISH BROWN SANDY LOAM IN THE UPPER PART, YELLOWISH BROWN SANDY CLAY LOAM IN THE MIDDLE PART AND
MOTTLED BROWNISH YELLOW, STRONG BROWN, YELLOWISH RED SANDY CLAY LOAM IN THE LOWER PART. SLOPES ARE 0 TO 10 PERCENT.

ESTIMATED SOIL PROPERTIES (A)

DEPTH (IN.)	USDA TEXTURE	UNIFIED	AASHTO	FRACT >3 IN (PCT)	PERCENT OF MATERIAL LESS THAN 3" PASSING SIEVE NO. 4	10	40	200	LIQUID LIMIT	PLAS-TICITY INDEX
0-17	SL, FSL	SM, SM-SC, SC	A-2	0	95-100	95-100	50-91	15-33	<25	NP-14
0-17	LS, LFS	SM	A-2	0	95-100	92-100	50-91	13-30	<20	NP
17-38	S., SCL, CL	SC, SM-SC, CL, CL-ML	A-2, A-4, A-6	0	95-100	91-100	70-96	30-55	20-38	4-15
38-70	SCL, CL, SC	SC, SM-SC, CL, CL-ML	A-4, A-6	0	100	98-100	65-98	36-72	20-45	4-22
70-99	VAR									

DEPTH (IN.)	CLAY (PCT <2MM)	MOIST BULK DENSITY (G/CM3)	PERMEA-BILITY (IN/HR)	AVAILABLE WATER CAPACITY (IN/IN)	SOIL REACTION (PH)	SALINITY (MMHOS/CM)	SHRINK-SWELL POTENTIAL	EROSION FACTORS K	T	WIND EROD. GROUP	ORGANIC MATTER (PCT)	CORROSIVITY STEEL	CONCRETE
0-17	5-20	1.3-1.6	2.0-6.0	0.06-0.10	4.5-6.0	–	LOW	.17	5	–	.5-2	MODERATE	HIGH
0-17	2-10	1.4-1.7	6.0-20	0.06-0.10	4.5-6.0	–	LOW	.17	5	2	.5-2		
17-38	18-35	1.3-1.6	0.6-2.0	0.10-0.15	4.5-5.5	–	LOW	.24					
38-70	20-40	1.2-1.5	0.6-2.0	0.10-0.15	4.5-5.5	–	LOW	.24					
70-99													

FLOODING			HIGH WATER TABLE			CEMENTED PAN		BEDROCK		SUBSIDENCE		HYD GRP	POTENT'L FROST ACTION
FREQUENCY	DURATION	MONTHS	DEPTH (FT)	KIND	MONTHS	DEPTH (IN)	HARDNESS	DEPTH (IN)	HARDNESS	INIT. (IN)	TOTAL (IN)		
NONE			4.0-6.0	APPARENT	JAN-MAR	–		≥60		–		B	–

SANITARY FACILITIES (B)

			CONSTRUCTION MATERIAL (B)	
SEPTIC TANK ABSORPTION FIELDS	0-8%: MODERATE-WETNESS 8-10%: MODERATE-SLOPE,WETNESS		ROADFILL	GOOD
SEWAGE LAGOON AREAS	0-2%: MODERATE-SEEPAGE,WETNESS 2-7%: MODERATE-SLOPE,SEEPAGE 7-10%: SEVERE-SLOPE		SAND	POOR-EXCESS FINES
SANITARY LANDFILL (TRENCH)	0-8%: MODERATE-WETNESS 8-10%: MODERATE-SLOPE		GRAVEL	UNSUITED-EXCESS FINES
SANITARY LANDFILL (AREA)	0-8%: MODERATE-WETNESS 8-10%: MODERATE-SLOPE		TOPSOIL	0-8% SL,FSL: GOOD 8-10% SL,FSL: FAIR-SLOPE 0-8% LS,LFS: FAIR-TOO SANDY 8-10% LS,LFS: FAIR-TOO SANDY,SLOPE
DAILY COVER FOR LANDFILL	0-8%: GOOD 8-10%: FAIR-SLOPE			

			WATER MANAGEMENT	
			POND RESERVOIR AREA	MODERATE-SEEPAGE

BUILDING SITE DEVELOPMENT (B)

SHALLOW EXCAVATIONS	0-8%: MODERATE-WETNESS 8-10%: MODERATE-SLOPE		EMBANKMENTS DIKES AND LEVEES	SLIGHT
DWELLINGS WITHOUT BASEMENTS	0-8%: SLIGHT 8-10%: MODERATE-SLOPE		EXCAVATED PONDS AQUIFER FED	SEVERE-NO WATER
DWELLINGS WITH BASEMENTS	0-8%: MODERATE-WETNESS 8-10%: MODERATE-SLOPE		DRAINAGE	NOT NEEDED
SMALL COMMERCIAL BUILDINGS	0-4%: MODERATE-WETNESS 4-8%: MODERATE-SLOPE 8-10%: SEVERE-SLOPE		IRRIGATION	FAST INTAKE,SLOPE
LOCAL ROADS AND STREETS	0-8%: SLIGHT 8-10%: MODERATE-SLOPE		TERRACES AND DIVERSIONS	SLOPE
LAWNS, LANDSCAPING AND GOLF FAIRWAYS			GRASSED WATERWAYS	SLOPE

REGIONAL INTERPRETATIONS

TABLE 31 *(continued)*

NORFOLK SERIES   NC0037

RECREATIONAL DEVELOPMENT (C)

| CAMP AREAS | 0-8%: SLIGHT / 8-10%: MODERATE-SLOPE | | PLAYGROUNDS | 0-2% LS,LFS: MODERATE-TOO SANDY / 0-2% SL,FSL: SLIGHT / 2-6% LS,LFS: MODERATE-SLOPE,TOO SANDY / 2-6% SL,FSL: MODERATE-SLOPE / 6+%: SEVERE-SLOPE |
| PICNIC AREAS | 0-8% SL,FSL: SLIGHT / 8-10% SL,FSL: MODERATE-SLOPE / 0-8% LS,LFS: MODERATE-TOO SANDY / 8-10% LS,LFS: MODERATE-SLOPE,TOO SANDY | | PATHS AND TRAILS | LS,LFS: MODERATE-TOO SANDY / SL,FSL: SLIGHT |

CAPABILITY AND YIELDS PER ACRE OF CROPS AND PASTURE (HIGH LEVEL MANAGEMENT)

CLASS-DETERMINING PHASE	CAPA-BILITY	COTTON LINT (LBS)		TOBACCO (LBS)		CORN (BU)		SOYBEANS (BU)		WHEAT (BU)		PEANUTS (LBS)		PASTURE (AUM)	
		NIRR	IRR.	NIRR	IRR.	NIRR	IRR.	NIRR	IRR.	NIRR	IRR.	NIRR	IRR.	NIRR	IRR.
0-2%	1	700		3000		110		40		60		4000		10.5	
2-6%	2E	650		2900		100		35		55		3700		10.0	
6-10%	3E	600		2700		90		30		50		3300		9.5	

WOODLAND SUITABILITY (D)

CLASS-DETERMINING PHASE	ORD SYM	MANAGEMENT PROBLEMS					POTENTIAL PRODUCTIVITY		TREES TO PLANT
		EROSION HAZARD	EQUIP. LIMIT	SEEDLING MORT'Y.	WINDTH. HAZARD	PLANT COMPET.	COMMON TREES	SITE INDX	
ALL	20	SLIGHT	SLIGHT	SLIGHT			LOBLOLLY PINE / LONGLEAF PINE / SLASH PINE	86 * / 68 * / 86	SLASH PINE / LOBLOLLY PINE

WINDBREAKS

CLASS-DETERMIN'G PHASE	SPECIES	HT	SPECIES	HT	SPECIES	HT	SPECIES	HT
	NONE							

WILDLIFE HABITAT SUITABILITY (E)

CLASS-DETERMINING PHASE	POTENTIAL FOR HABITAT ELEMENTS									POTENTIAL AS HABITAT FOR:				
	GRAIN & SEED	GRASS & LEGUME	WILD HERB.	HARDWD TREES	CONIFER PLANTS	SHRUBS	WETLAND PLANTS	SHALLOW WATER	OPENLD WILDLF	WOODLD WILDLF	WETLAND WILDLF	RANGELD WILDLF
0-6%	GOOD	GOOD	GOOD	GOOD	GOOD	–	POOR	V. POOR	GOOD	GOOD	V. POOR	–
6-10%	FAIR	GOOD	GOOD	GOOD	GOOD	–	V. POOR	V. POOR	GOOD	GOOD	V. POOR	–

POTENTIAL NATIVE PLANT COMMUNITY (RANGELAND OR FOREST UNDERSTORY VEGETATION)

COMMON PLANT NAME	PLANT SYMBOL (NLSPN)	PERCENTAGE COMPOSITION (DRY WEIGHT) BY CLASS DETERMINING PHASE				
AMERICAN HOLLY	ILOP					
FLOWERING DOGWOOD	COFL2					
COMMON PERSIMMON	DIVI5					
BLUERIDGE BLUEBERRY	VAVA					
GREENBRIER	SMILA2					
POTENTIAL PRODUCTION (LBS./AC. DRY WT): FAVORABLE YEARS / NORMAL YEARS / UNFAVORABLE YEARS						

FOOTNOTES

A ESTIMATES OF ENGINEERING PROPERTIES BASED ON 2 PEDONS FROM ROBESON AND WAYNE COUNTIES, N.C.
B RATINGS BASED ON "GUIDE FOR INTERPRETING ENGINEERING USES OF SOILS". NOV. 1971.
1 RATINGS BASED ON SRWPC COMMITTEE 4 GUIDE. APRIL 1970.
C RECREATION RATINGS BASED ON SOILS MEMORANDUM-69. OCT. 1968.
D WOODLAND RATINGS BASED ON SOIL SURVEY INTERPRETATIONS FOR WOODLAND" PROGRESS REPORT W-16. 1-70.
E WILDLIFE RATINGS BASED ON SOILS MEMORANDUM-74. JAN. 1972.
* SITE INDEX IS A SUMMARY OF 5 OR MORE MEASUREMENTS ON THIS SOIL.

nities include holly, dogwood, persimmon, blueberry, and greenbrier. In short, the computer Form 5 gives a quick statement of the suitabilities and limitations of each soil for all the common uses of land in an area.

The computer format of soil survey interpretations in Tables 29-31 has been used very effectively in data printout books to compare soils for different uses, and for making computer maps of suitabilities and limitations of different areas. The computer format has also been used to accelerate the publication of soil survey reports, because the computer can be used to store the data and then print it out rapidly in the tabular form needed. Ratings of soils for sanitary facilities, for example, are entered into the form in Table 29 according to the standard computer language in Table 30. Ratings for soils in a survey area can then be quickly assembled from the output forms in Table 31, and the computer can also print the ratings in tables for direct publication in soil survey reports. Thus, the ultimate goal of the Cooperative Soil Survey is to make soil survey data and interpretations available for use as quickly as possible—in a form that will be most understandable to users. Many additional possibilities can be developed to provide better and wider use of this soil information through computers and other data manipulations.

CHAPTER 6

ENGINEERING APPLICATIONS

Engineering applications of soil surveys have contributed a great deal for making soil maps and descriptions more useful. As Table 29 illustrates, engineering analyses and data that make soil information more quantitative thereby provide a data base for soil profile descriptions so they can be more fully utilized. Engineering applications of soil information are fully as important for agricultural operations as for construction activities. Many soil improvements (drainage, grading, compaction, trafficability) involve soil character changes under stress or with changing moisture conditions whether for rural or urban developments.

Engineering and pedological data are compatible, and numerous correlations and rating systems have been devised to mesh the disciplines of engineering and soil science. Increasingly, soil survey workers have adopted and adapted engineering tests, to make pedology more quantitative. The Atterberg limits (liquid limit, plastic limit, plasticity index), for example, have been readily adopted into soil sampling and mapping programs, and the data are commonly published in soil survey reports. Some engineering tests, such as shrink-swell potential (Table 29), have been adapted as standard tests to help characterize all soils. Shrink-swell potential determines volume change with moisture differences in soils, determined from the coefficient of linear extensibility (COLE). The COLE is determined by the vertical component of swelling of a natural soil clod:

$$\frac{L_m - L_d}{L_d}$$

where L_m = length of moist sample
L_d = length of dry sample

Shrink-swell interpretations of soils affect planning for such structures as foundations, roads, embankments, and canals. Low shrink-swell potential has COLE values < 0.03, moderate 0.03-0.06, and high > 0.06 COLE values.

Soil movement under varying moisture and temperature conditions is illustrated in Figure 42. Due to soil movement, foundations are fractured, alfalfa roots are disrupted and plants die, landslides are initiated on steep slopes, and trafficability by wheeled vehicles becomes difficult. Corrective measures need engineering solutions to provide drainage and control runoff not only at the exact site in question but also on surrounding soils both upslope and downslope.

All the soil factors tabulated and summarized in Table 29 are important and have been quantified to a considerable extent. Steel corrosivity, for example, is caused by oxidation (rusting) of iron buried in different soils, and is influenced by soil acidity,

FIGURE 42 *Secondary road on soils high in clays and silts with a critical plasticity index. The photograph was taken in early spring in Ithaca, New York, after frost action had broken up much of the road.*

texture, drainage, and conductivity (FAO, 1977). Soils with low corrosivity generally have acidity below 8.0 meq/100 g of soil, and conductivity of the saturation extract less than 0.2 mmho per centimeter at 25°C. Soils with moderate corrosivity have total acidity of 8.0–12.0 meq/100 g of soil, and conductivity of 0.2–0.4 mmho. Soils with high corrosivity have total acidity greater than 12.0 meq/100g of soil and electrical conductivity greater than 0.4 mmho. Wet soils with fine textures and high salt content are generally more corrosive to uncoated steel than are well-drained coarser soils.

Concrete corrosivity in soils (Table 29) is influenced by sulfates, texture, and acidity. Soils with low corrosivity for concrete generally include coarse-textured soils, organic soils that have pH greater than 6.5, medium- and fine-textured soils that have a pH greater than 6.0, and soils that contain less than 1,000 parts per million of water-soluble sulfate (as SO_4^{2-}). Soils with moderate corrosivity for concrete generally include coarse-textured soils and organic soils that have a pH of 5.5–6.5, medium and fine textured soils that have a pH of 5.0–6.0, and soils that contain 1,000–7,000 parts per million of water-soluble sulfate. Soils with high corrosivity for concrete include coarse-textured soils and organic soils that have a pH of 5.5 or less, medium- and fine-textured soils that have a pH of 5.0 or less, and soils that contain more than 7,000 parts per million of water-soluble sulfate.

Ratings of soils for engineering purposes (Tables 29–31) can conveniently be made following the format illustrated in Tables 17 and 26. Consideration of soil map units as potential sources of sand and gravel for construction purposes (Table 32) is largely an

TABLE 32 *Ratings of soils as sources of aggregate (sand and gravel) material for construction purposes (adapted from FAO, 1977; and Soil Survey Staff, 1971)*

Probable source		Improbable source	
Good	Moderate	Poor	Unsuited
SW	SW–SM	SM	All other groups in the
SP	SP–SM	SW–SC	Unified soil
		SP–SC	classification system
GW	GP–GM		
GP	GW–GM	GM	
		GP–GC	
		GW–GC	

interpretation of the Unified soil groups (Table 16). Soil map units with horizons classified SW and SP are probably good sources of sand, and GW and GP soils are probably good sources of gravel. Areas likely to be moderate in suitability include SW-SM, SP-SM, GP-GM, and GW-GM. Poor sources include soils classified as SM, SW-SC, SP-SC, GM, GP-GC, and GW-GC. All the other Unified soil groups are likely to be unsuitable soil map units for sand and gravel mining operations.

COMMUNITY DEVELOPMENT

Criteria for making ratings of soil map units for secondary roads are listed in Table 33. Slope, of course, is a crucial criterion. Hard bedrock makes excavation difficult and

TABLE 33 *Ratings of soils for secondary roads (adapted from FAO, 1977; and Soil Survey Staff, 1971)*

	Soil limitation rating		
Item affecting use	Slight	Moderate	Severe
Slope	0–8%	8–15%	> 15%
Depth to hard bedrock	> 40 in.	20–40 in.	< 20 in.
Unified soil group for subgrade	GW, GP, SW, SP, GM, GC, SM, SC	CL with PI < 15	CL with PI > 15, CH, MH, OH, OL, Pt
Shrink–swell potential	Low	Moderate	High
Susceptibility to frost action	Low	Moderate	High
Stoniness	Few	Some	Many
Soil drainage class	Well drained, moderately well drained	Somewhat poorly drained	Poorly drained, very poorly drained
Flooding frequency	None	Occasionally (soils flooded less than once in 5 years)	Frequently (soils flooded more than once in 5 years)

TABLE 34 *Ratings of soils for buildings without basements (adapted from FAO, 1977; and Soil Survey Staff, 1971)*

Item affecting use	Soil limitation rating		
	Slight	Moderate	Severe
Slope	0–8%	8–15%	> 15%
Shrink–swell potential	Low	Moderate	High
Unified soil group	GW, GP, SW, SP, GM, GC, SM, SC, CL with plasticity index < 15	ML, CL with plasticity index > 15	CH, MH, OL, OH
Potential frost action	Low	Moderate	High
Stones and rocks	None	Some	Many
Flooding	None	None	Rare, occasional, or frequent
Depth to bedrock	> 40 in.	20–40 in.	< 20 in.
Soil drainage class	Well drained, moderately well drained	Somewhat poorly drained	Poorly drained, very poorly drained

expensive. Subgrade conditions are determined largely by the Unified soil groups. Shrink-swell and frost heaving should be low to moderate or roadbeds will be difficult to maintain (Fig. 42). Stones interfere with excavation and grading. Wet soils cause difficulties with shrink–swell and frost heaving, and drainage or fill in wet areas is expensive. Flooding is a hazard, when sections of roads wash out or are covered with water even for brief periods. The criteria and classes, of course, can be refined if necessary. More details on the ratings are given in the references cited.

For building houses and other structures, ratings have been made (Table 34) to show

TABLE 35 *Ratings of soils for effluent disposal in septic tank seepage fields (adapted from FAO, 1977; and Soil Survey Staff, 1971)*

Item affecting use	Soil limitation rating		
	Slight	Moderate	Severe
Permeability class	Rapid, moderately rapid	Moderate	Moderately slow, slow
Water conductivity rate	> 1.0 in./hr	0.6–1.0 in./hr	< 0.6 in./hr
Percolation rate in auger hole	Faster than 45 min/in.	45–60 min/in.	Slower than 60 min/in.
Soil drainage class	Well drained, moderately well drained	Somewhat poorly drained	Poorly drained, very poorly drained
Flooding	None	Rare	Occasional or frequent
Slope	0–8%	8–15%	> 15%
Depth to hard rock or impermeable layer	> 72 in.	48–72 in.	< 48 in.
Stoniness and rockiness	None	Some	Many

limitations and suitabilities for foundations. Nearly level slopes reduce the costs for site leveling. Shrink-swell and frost action in soil movement should be low. Unified soil groups with good bearing and compaction characteristics are better. Stoniness and rockiness cause many excavation problems in some soils. Flooding and wet soils should be avoided, if possible. These ratings do not replace or eliminate the need for on-site investigation, but the ratings do help the engineer or contractor to determine the likely soil problems even before construction starts, and to plan for the improvement of the site as the project is implemented. Sites rated "severe" probably have problems of such great magnitude that construction may not be feasible on those soil map units.

A home builder can use the soil map to determine road location to the site, nature of footings required for the foundation, and septic tank seepage layout in addition to using soil information in landscape design around the house. Soil map unit and geomorphology information can also help the contractor to predict likely well yields from aquifers beneath the soils. Each user of soil maps can exploit the information in numerous ways, but the soil maps are most useful if the user has extensive knowledge about the maps and how the soils were described.

WASTE DISPOSAL

Many waste disposal problems can be solved by wise use of soil resources. Table 35 gives the criteria for ratings of soils for effluent disposal in septic tank seepage fields—where public sewerage facilities are not available.

Septic tank effluent disposal depends upon soil permeability to water or effluent, the conductivity rate, and the percolation rate; rapid rates are more suitable than slower rates, and pollution is generally not a problem if houses and effluent disposal systems are not closely spaced. Poorly and very poorly drained soils have severe limitations for effluent disposal because the high water table prevents the contamination effluent from seeping away. Flooding spreads pollution when the effluent seeps to the soil surface, and can be a serious health hazard. Steep slopes cause problems for construction, maintenance, and operation of seepage fields—and landslides may be a problem on some soils. Stones and bedrock, of course, are impermeable and cause problems during construction, especially.

Sewage lagoons (Table 36) are being used increasingly for disposal of wastes from small communities and for some industrial disposal systems. Aerobic sewage lagoons or waste stabilization ponds are shallow ponds that hold waste liquids for the time necessary for microbial decomposition. Sewage lagoons require considerations of the soils as a vessel for the impounded liquid, and as soil material for the enclosing embankment. Sewage lagoons, of course, must be impermeable or only slowly permeable, without large amounts of seepage. Bedrock causes problems in excavation. Sloping soil map units are undesirable, because large fill and embankment costs are involved where slopes must be leveled. Soils containing moderate to high amounts of organic matter are unsuitable for the basin floor even if the floor is underlain by suitable soil material. Organic matter promotes growth of aquatic plants which are detrimental to proper functioning of the lagoon. Sometimes liquids are pumped out of lagoons after treatment periods and sprayed onto adjacent soils. In time, solids also accumulate in the bottoms of sewage lagoons and must be removed. Soil survey interpretations can also be used to determine suitable areas for irrigation of liquids and spreading of the remaining solid sludge.

TABLE 36 *Ratings of soils for waste disposal in sewage lagoons (adapted from FAO, 1977; and Soil Survey Staff, 1971)*

Item affecting use	Soil limitation rating		
	Slight	Moderate	Severe
Depth to permanent or fluctuating water table	> 60 in.	40–60 in.	< 40 in.
Permeability	0.6 in./hr	0.6–2.0 in./hr	> 2.0 in./hr
Depth to bedrock	> 60 in.	40–60 in.	< 40 in.
Slope	< 2%	2–7%	> 7%
Coarse fragments < 10 in. diameter, % by volume	< 20%	20–50%	> 50%
Percent of soil surface covered by coarse fragments < 10 in. diameter	< 3%	3–15%	> 15%
Organic matter	< 2%	2–15%	> 15%
Flooding	None	None	Soils subject to flooding
Unified soil groups	CG, SC, CL, and CH	GM, ML, SM, and MH	GP, GW, SW, SP, OL, OH, and Pt

Most communities have solid waste disposal problems. Table 37 outlines the soil characteristics that are important for landfill burial in trenches. Sanitary landfill in trenches consists of dug or bulldozed sections in which refuse is buried and covered with a thin layer of soil material at least 6 in. thick at least daily, and compacted. Wet soils with high water tables with clayey textures are not good; rapidly permeable soils also

TABLE 37 *Ratings of soils for solid waste disposal in sanitary landfill trenches (adapted from FAO, 1977; and Soil Survey Staff, 1971)*

Item affecting use	Soil limitation rating		
	Slight	Moderate	Severe
Depth to seasonal water table	> 72 in.	> 72 in.	< 72 in.
Soil drainage class	Well drained, moderately well drained	Somewhat poorly drained	Poorly drained, very poorly drained
Flooding	None	Rare	Occasional or frequent
Permeability	Slower than 2.0 in./hr	Slower than 2.0 in./hr	Faster than 2.0 in./hr
Slope	< 15%	15–25%	> 25%
Dominant soil texture to a depth of 60 in.	sl, l, sil, scl	sicl, cl, sc, ls	sic, c, muck, peat, gravel and sand
Depth to hard bedrock	> 72 in.	> 72 in.	< 72 in.
Depth to rippable bedrock	> 60 in.	< 60 in.	< 60 in.
Stones and rocks	None	Few	Some or many

provide considerable risks for groundwater pollution. Rockiness and stoniness cause difficulties for the excavations and compactions. A final cover of soil material at least 2 ft thick is placed over the landfill when the trenches are full. For cover material (Table 29), soils with very friable and friable consistence are good, soils with loose and firm consistence are fair, and soils with very firm and extremely firm consistence are poor. Good soil textures for cover material include sandy loam, loam, silt loam, and sandy clay loam; fair textures include silty clay, clay, muck, peat, and sand. Thicker soil profile material is more desirable if excavation and hauling is to take place. Most landfills have effluent seepage downslope, especially in humid climates, so that soil maps should be used to plan impounding structures, water diversions, drainage, and irrigation of upslope and downslope areas to reduce pollution of the environment.

In cities, a major problem is disposal of sewage sludge. Table 38 lists criteria for rating soils around cities, and for disposal of biodegradable solids. The best soils for spreading or injection of sewage solids have moderate permeability, good drainage, slow runoff, high water-holding capacity, and no flooding hazards. Soils to be avoided for waste applications are those with very rapid or very slow permeability, excessively good or very poor drainage, rapid runoff, frequent flooding, and low water-holding capacity for vegetative growth. These rating tables represent guidelines, and are to be used with refinements to fit each local soil site condition. With fullest use of soil maps and other soil information, however, many rural and urban problems can be solved and many engineering feasibility determinations can be made in the project planning stages. For large investments, special soil maps at large scale and special soils investigations can be made to give highly detailed information that is more reliable than routine soil mapping procedures. When soils data are designed to fit the environment and the development problem, engineering applications of soil information can be made in harmony with the natural environment to give a more efficient and aesthetic habitat for human beings and other living things.

TABLE 38 *Ratings of soils for application of sewage sludge (adapted from Loehr, 1977)*

Item affecting use	Soil limitation rating		
	Slight	Moderate	Severe
Permeability of the most restrictive layer above 60 in.	0.6–6.0 in./hr	6–20 and 0.2–0.6 in./hr	> 20 and < 0.2 in./hr
Soil drainage class	Well drained, moderately well drained	Somewhat excessively drained, somewhat poorly drained	Excessively drained, poorly drained, very poorly drained
Runoff	Ponded, very slow, slow	Medium	Rapid, very rapid
Flooding	None	None	Rare or frequent
Available water capacity from 0.60 in. or to a root-limiting layer	> 8 in. (humid regions) > 3 in. (arid regions)	3–8 in. (humid regions) Moderate class not used in arid regions	< 3 in. (humid regions) < 3 in. (arid regions)

CHAPTER 7

AGRICULTURAL LAND CLASSIFICATION

Agricultural production is vital to provide the subsistence base for every civilization and society. Land classification is the grouping of soil map units "primarily on the basis of their capability to produce common cultivated crops and pasture plants without deterioration over a long period of time" (Klingebiel and Montgomery, 1966). The land capability classification is based on the aggregation of soil map areas, and the units of classification are published for each soil map unit in modern soil survey reports. A bulletin and set of 50 color slides is available to introduce the concepts of the classification in a format readily understood by laypersons (SCS, 1969).

LAND CAPABILITY

Capability units, into which soil mapping units are grouped, have similar potentials and continuing limitations. Soil map units put into a capability unit are sufficiently uniform to produce similar kinds of cultivated crops and pasture plants, require similar conservation treatment and management, and have comparable potential productivity. Use of capability units condenses and simplifies soil map unit information for planning and management purposes in fields on farms. A capability unit is designated by a symbol such as IIIe-2. The Roman numeral designates the capability class of lands that have the same relative degree of limitation; the risks of soil damage or limitation in use become progressively greater from Class I to Class VIII. The lowercase letters designate subclasses that have the same major conservation problem: e (erosion and runoff), w (excess water), s (root zone limitations), and c (climatic limitations). The Arabic numbers indicate the capability unit (subdivision) within each capability class and subclass.

Classes I, II, III, and IV are generally considered to be land suited to cultivation (Fig. 43). Classes V, VI, VII, and VIII are generally not suited to cultivation and are limited in use (Fig. 44). With major earth-moving or other costly reclamation, certain soils in Classes V, VI, VII, and VIII may be made fit for cropping.

Briefly, the land capability classification definitions (Klingebiel and Montgomery, 1966) are:

Class I   Soils with few limitations that restrict their use
Class II   Soils with some limitations that reduce the choice of plants or require moderate conservation practices
Class III   Soils with severe limitations that reduce the choice of plants or require special conservation practices

FIGURE 43 *Class I land near Phoenix, Arizona. These soils are deep, well drained, and easily worked. They hold water well and are responsive to inputs of fertilizer. They are nearly level and have low erosion. Class I land is suited to a wide range of plants and may be safely used for cultivated crops, pasture, rangeland, woodland, and wildlife.*

FIGURE 44 *Class VIII land near Phoenix, Arizona. These soils have limitations that preclude their use for commercial plant production and restrict their use to recreation, wildlife, water supply, or to aesthetic purposes. Class VIII land cannot be expected to return significant benefits from management for crops, grasses, or trees.*

Class IV   Soils with very severe limitations that restrict the choice of plants or require very careful management

Class V   Soils with little or no erosion hazards but with other limitations that restrict their use largely to pasture, rangeland, woodland, or wildlife food and cover

Class VI   Soils with severe limitations that make them generally unsuited to cultivation and restrict their use largely to pasture, rangeland, woodland, or wildlife food and cover

Class VII   Soils with very severe limitations that make them unsuited to cultivation and that restrict their use largely to grazing, woodland, or wildlife

Class VIII Soils with limitations that preclude their use for commercial plant production and restrict their use to recreation, wildlife, water supply, or to aesthetic purposes

Land classifications can be changed if soils are improved, and relatively permanent amelioration projects will greatly enhance the productivity and management of soil areas. Dry soils can be irrigated, stony soils can be cleared of coarse fragments, salty soils can be leached and chemically treated, wet soils can be drained, and alluvial soils can be protected from flooding. Soil and water conservation plans made by the Soil Conservation Service clearly illustrate how land management systems can be tailored to fit the needs of each farmer or rancher in accord with the different soil resources in each field of each farm. Brochures are available (SCS, 1973) to show how land capability classes permit individual soil areas to be best managed. The principles of land classification can be applied equally well to both rural and urban areas; some of the techniques may be different but the principles are similar to help solve every land management problem.

A publication by Brown (1963) illustrates how the land capability classification can help landowners and others use soil maps. Soils of the Flat Top Ranch in Bosque County, Texas, were grouped into capability classes, and the minimum conservation treatment was specified for each class to meet the management objectives of the landowner. Deep-rooted legumes or perennial grasses were to be grown one out of every four years on Class I soils, one out of every three years on Class II soils, one out of every two years on certain Class III soils, and four out of five years on Class IV soils. Ranch grasslands in the higher capability classes but not suitable for farming were to be managed for better root development and plant vigor by leaving at least half of the annual growth and by providing timely rest periods to maintain vigor of desirable adapted plants.

Different land classification systems used in various countries have been reviewed and summarized by the Food and Agricultural Organization of the United Nations (FAO, 1974). The land capability classification (Klingebiel and Montgomery, 1966) is widely used and has been adapted in many countries. Countries without detailed soil mapping programs cannot use soil maps to help plan operation of single farms, so planning must be done in these countries with more general soil maps. Very large scale soil maps are commonly used for land classification in irrigation project areas; operations such as land leveling require highly detailed soil information for optimum soil improvement. Land-use mapping is also common in many places, and computerization of soil rating systems is being increasingly used for tax assessments and many other purposes.

PRODUCTIVITY INDEX

In Hawaii (Nelson et al., 1963) ratings were developed for soil map units based on general character of the soil profile, texture of the surface soil, slope of the topography, climate, and other physical conditions affecting use of the land. A land productivity index was used to calculate percentage values appropriate for local conditions:

$$\text{land productivity index} = A \times B \times C \times X \times Y$$

where A = percentage rating for the general character of the soil profile
$\quad$ B = percentage rating for the texture of the surface horizon
$\quad$ C = percentage rating for the slope of the land
$\quad$ X = percentage rating for site conditions other than those covered in factors A
$\qquad$ B, and C (e.g., salinity, soil reaction, freedom from damaging winds)
$\quad$ Y = percentage rating for rainfall

LAND ECONOMICS

In New York State, farm areas were mapped by land economists using aerial photographs and topographic maps at a scale of 1:24,000. Evaluation of the farm areas was done in a fairly systematic and quantitative fashion, mainly from the economic point of view to evaluate capital investment and potentialities, but considering all other relevant information as well. Maps were made of the farms in the following fashion:

A crew of four people traveling in a carryall type of vehicle made the field examinations (Nobe et al., 1960). The vehicle was equipped with a worktable, filing cabinets, and a recording machine. One person followed the route on the aerial photographs as the group traveled, and made continuous verbal notes on the recording machine. These notes described the buildings, fields, crops, livestock, and other visible characteristics of the farms. Points were numbered on the air photos and the notes were made with reference to those numbers to permit an interwoven record. Another person, following the route on topographic and soil maps, made a written record of observations about relationships between the maps and the kind and level of farming observed. The third person supplied the others with photographs and maps from the files, and consulted reports of soil studies, farm management surveys, climatic data, and other specialized materials as these became relevant at various points along the route. The fourth person was the driver. The crew made north–south traverses across the entire area at 5- to 10-mile intervals, traveling in as nearly straight lines as possible. In this way, they traveled on all types of roads and saw as wide a range as possible of the use conditions that exist in the area.

The air photos (Nobe et al., 1960), for both the areas seen in the field and those not seen, were later examined in detail in the office. All notes made in the field were used at this time. Studies of the photographs for areas that had been seen, together with the notes, made it possible to classify by aerial photographic interpretation the areas that had not been seen. Further studies of soil maps, climate data, farm management surveys, and other materials were made. This work produced a classification of farms; a classification arranged according to the intensities with which farms are being used and the likely viability of those areas to continue in agricultural production in the future. Farms were classified individually, then the areas were mapped to show differences among farms and between regions.

TABLE 39 *Areas in farms and relative agricultural prosperity in some of the largest soil associations in five counties around Syracuse, New York (adapted from Olson et al., 1969)*

| Soil association (1) | Area (mi²) and intensity of use for farming | | | Prosperity index (5) |
	High (2)	Medium (3)	Low (4)	Ratio (2)/(4)
Palmyra	85	20	5	17.0
Honeoye–Lima	318	87	27	11.8
Lansing–Conesus	49	38	9	5.4
Ontario–Hilton	109	82	22	5.0
Langford–Erie	43	44	24	1.8
Muck–peat	16	14	13	1.2
Langford-Erie-Lordstown	28	75	32	0.9
Volusia-Mardin-Lordstown	24	45	30	0.8
Lordstown-Volusia-Mardin	38	56	69	0.6
Minoa-Lamson-Canandaigua	6	25	30	0.2
Collamer–Niagara	3	17	26	0.1
Sodus–Ira	11	51	98	0.1
Ira–Sodus	2	26	51	0.0
Worth-Empeyville	0	9	37	0.0
Total	732	589	473	1.5

Areas used at a high intensity have farms that appear capable of supporting viable farm businesses for the foreseeable future. Their land is well adapted to modern farming methods, current capital investments are usually adequate and in good repair, new improvements usually keep pace with technical developments, and most operators are skillful and dedicated to continuing their units in agriculture.

Areas used at a medium level of intensity have farms near enough to the economic margin to make their future somewhat uncertain. Income prospects now provide these farm families the option of continuing in farming, but not all will prefer this option to nonfarm employment, and further developments in farm technology will tend to put these operators gradually at a disadvantage. Only two-thirds of these farms are expected to pass to the next generation as full-time units, but individual circumstances will determine which businesses discontinue.

Areas used at a low level of intensity have farms judged obsolete for full-time use under modern farming conditions. These include full-time and part-time farms but not units converted primarily to residences or other nonfarm purposes (Nobe et al., 1960).

After the farm areas were mapped, area boundaries were generalized and maps were reduced and redrawn at a scale of 1:250,000. Soil association maps were then compared (Olson et al., 1969) with the farmland intensity (or viability) maps by overlay techniques. Table 39 is a summary of some of the area comparisons. When areas of soils are compared with areas of farmland used at high and low levels of intensity, a ratio of high/low intensity can be used to give a "prosperity index" for farming purposes. When the "prosperity index" is compared for each soil, the ranking is remarkably in accord with the ranking of the soil characteristics as given in Table 27. Thus, the current prosperity of farms is shown to be closely related to soil characteristics, and the future productivity and prosperity of the land is also likely to be determined primarily by the soil properties.

LAND USE

Figure 45 is a more detailed example of relating land use to soil map units. In this area, more than 75 percent of the active pasture and cropland has slopes of less than 10 percent, but only about 20 percent of the abandoned farmland has slopes of less than 10 percent. About 40 percent of the forest land has 35–70 percent slopes. Much of the pasture land (about 25%) has soils with seasonal water tables less than 5 in. below the

FIGURE 45   *Detailed soil map from part of Map Sheet 31 of soil survey of Tompkins County, New York (Neeley et al., 1965), from which a computer analysis was made of relationships between soils and land use. Most of the nearly level, farmed valley has soils formed in alluvium (Em, Mo) and glacial outwash (CnB, HdA); forested soils on steep slopes (BtF, LoF) are formed in glacial till only moderately deep to bedrock in places; soils on upper slopes (MaC, VbB) have seasonal seepage water at shallow depths in the glacial till.*

surface. Most of the cropland (about 60 percent) has soils with moderate to rapid permeability. More than 40 percent of the inactive farmland and forests has soils with fragipans, but only about 10 percent of the active cropland has such soils. Obviously, soils have played a major role in land development in the past, have influenced trends in land-use shifts, and will play prominent roles in land use in the future.

Land classification and soil classification are rapidly improving and being put to greater use. Temperature and moisture seasonal changes in soils are being increasingly quantified. Some of the soil temperature classes identified by the Cooperative Soil Survey, for example, are:

Pergelic	Mean annual soil temperature less than $0°C$ at 20 in.
Cryic	Mean annual soil temperature more than $0°C$, but less than $8°C$
Frigid	Mean annual soil temperature less than $8°C$, and more than $5°C$ difference between mean winter and mean summer soil temperatures
Mesic	Mean annual soil temperature $8°C$ or more, but less than $15°C$, and more than $5°C$ difference between mean summer and mean winter soil temperature
Thermic	Mean annual soil temperature $15°C$ or more, but less than $22°C$, and more than $5°C$ difference between mean summer and mean winter soil temperatures
Hyperthermic	Mean annual soil temperature $22°C$ or more, and more than $5°C$ difference between mean summer and mean winter soil temperatures

Such temperature classes have obvious implications for agriculture and growth of plants in different soil regions, and they also have certain engineering applications. Even further refinements in the classes will probably be needed by land classification experts for planning of many projects. Soil moisture regimes, as affected by climate, are also receiving prominent places in soil and land classification systems.

CHAPTER 8

EROSION
CONTROL

Soil erosion is probably the most destructive process that acts to reduce production from the land. Soil surveys are used primarily for making management decisions, and many decisions about land use involve erosion control. Topsoils and A horizons generally contain the most nutrients and best structure for plant growth, and any materials eroded from the upper part of the soil profile have a detrimental effect upon crop yields and plant growth. Soil erosion is fully as important in urban areas as in rural areas, and is usually associated with runoff and sedimentation, which can cause great damage downslope as well as on the higher elevations.

The reason for making detailed soil maps accurate enough to show soil differences down to 1 hectare in size is to have soil resource inventories that can be used for practical planning purposes on farms or in project development areas. Figure 46 illustrates how soil differences within fields are critical for making management decisions, and why it is important to design cropping systems in accord with the soil capabilities and limitations. The soils on the upper slopes in Figure 46 need to have good crop cover, contour cultivation, strip cropping, grass cover, and terraces to reduce erosion and prevent flooding and sedimentation on the lower slopes. The erosion and runoff problem is cumulative, of course, because each farm contributes runoff and sediment to the next farm downslope, until flooding may reach disastrous proportions even in urban areas within drainage basins. With good soil management, soil loss and runoff can be reduced, productivity can be maintained, and flooding controlled or prevented.

Control of erosion problems is a complex matter, and involves such activities as educational programs, engineering, and agronomic techniques. Often, erosion control is expensive, so that individual farmers cannot afford to build the terraces or construct the waterways that are required; in these cases, governmental support is essential. The concept of the "tragedy of the commons" (Hardin, 1968) applies very well to soil management and erosion control. Where erosion is serious and control is expensive, farmers on their own are likely to operate in the short run to maximize returns and allow their soils to erode. Soil erosion, of course, is usually an insidious process that may be hardly noticeable for several years. Ultimately, however, the productivity of the soil declines abruptly when the soils erode down to the bedrock, hardpan, or less fertile subsoils. The "tragedy of the commons" for soil erosion is well illustrated in many areas in Latin America, Africa, and Asia where farmers are too poor to control erosion over the long run, and where governmental programs for erosion control have been ineffective. Eckholm (1976) has excellently documented the environmental stress and world food prospects as influenced by the damaging effects of soil erosion.

FIGURE 46 *Eroded farm near Walla Walla, Washington, after a heavy rain. The erosive loess soils are readily damaged by erosion during rainstorms, and sediments are also detrimental downslope. This view excellently illustrates how different conservation practices are needed on different soils even within the same field. Conservation practices must be designed in accord with the characteristics of the different soils to be most effective (photo by Soil Conservation Service).*

Figures 47 and 48 illustrate typical soil erosion problems that arise when soils are used and abused beyond their capacity. In Figure 47, steep erosive soils were cleared of the forest and used for cropping where they should have been carefully protected under forest; probably both excessive human and animal populations contributed to the destruction of the forest lands. Eventually, sheet erosion and gullies made these soils almost useless for cropping, but still they continue to erode as they are overgrazed and overexploited. Reforestation would be a good start toward control of this erosion, but

FIGURE 47 *Eroded fields near Apartaderos in northwestern Venezuela. The horizontal field boundaries on these steep slopes indicate that the gullies were probably induced and accelerated by human activities on the land. The remnant trees visible in the photo indicate that the natural forest would be likely to return in a few years if selected areas were fenced off and not used.*

FIGURE 48 *Sediments in the Maracaibo basin near El Vigia in northwestern Venezuela. Soil erosion from uplands like those shown in Figure 47 damages the lowlands also. Such sediments are difficult to remove and will increase flooding and raise water tables if the channel is not kept open. Increased investments must be made in the uplands as well as in the lowlands to assure lasting agricultural productivity.*

the large human populations and the grazing animals in the area make revegetation unlikely.

Downstream sediments also cause damage in the lowlands, as Figure 48 illustrates. Ecological and environmental abuse in the uplands have direct influence on the lowlands —which often may be at a considerable distance from the source of the erosion problem. Sediments commonly clog drainage channels and cause increased flooding and other problems. The magnitude and interrelatedness of the erosion problems demand that society and government have a hand in the solution of the dilemma, because individual farmers by themselves cannot completely solve all the erosion problems. People working collectively, with common goals, can better work to solve the soil erosion problems when educational programs and capital for investment in erosion control structures are available.

Many techniques can be used to reduce and control soil erosion (Fig. 49). Contour cultivation and strip-cropping are usually relatively inexpensive. Terraces are more expensive. Often crops can be grown which will provide greater vegetative cover and protection for the soil surface from erosion. Sod crops or pasture can be put on steep slopes, and trees can be planted in areas already seriously eroded. The ecology in many areas is in a delicate balance, and any disruptions should be carefully monitored and managed.

SOIL-LOSS EQUATION

Fortunately, a great deal of work has been done to measure soil erosion on different soils and to devise techniques to better control soil losses. Wischmeier and Smith (1978) have summarized factors for a soil-loss equation which permits planners and farmers to

FIGURE 49 *Contour cultivation on erosive sandy soils southwest of Brasilia. These soils erode very easily, but erosion can be at least partly controlled by terraces and farming on the contour. Erosion control could be even further improved by sod crops and pasture, strip-cropping, and more protective mulch at the soil surface.*

predict the average rate of soil erosion for each feasible alternative combination of crop system and management practices in association with a specific soil map unit, rainfall pattern, and topography. Virtually all of the work of the Soil Conservation Service in the United States is tied to the detailed soil maps and the soil-loss-equation factors for recommending management systems on different kinds of soils. Cooperative research projects at more than 50 locations contributed more than 10,000 plot-years of basic runoff and soil loss data over more than 50 years to permit summarizing and statistical analyses of empirical data by computers. The measurements made under actual field conditions have been supplemented by experimental studies with artificial rainmakers under both field and laboratory conditions. Based on soil profile characteristics, tolerable soil loss values have been determined that will permit the planning of agricultural systems for sustained production from each soil map unit.

The soil-loss equation (Wischmeier and Smith, 1978) is expressed as

$$A = RKLSCP$$

where *A* is the computed soil loss per unit area, expressed in the units selected for *K* and for the period selected for *R*; in practice, these are usually so selected that they compute *A* in tons per acre per year, but other units can be selected

    R, the rainfall and runoff factor, is the number of rainfall erosion index units, plus a factor for runoff from snowmelt or applied water where such runoff is significant

    K, the soil erodibility factor, is the soil loss rate per erosion index unit for a specified soil as measured on a unit plot, which is defined as a 72.6-ft length of uniform 9 percent slope continuously in clean tilled fallow

    L, the slope length factor, is the ratio of soil loss from the field slope length to that from a 72.6-ft length under identical conditions

    S, the slope steepness factor, is the ratio of soil loss from the field slope gradient to that from a 9 percent slope under otherwise identical conditions

    C, the cover and management factor, is the ratio of soil loss from an area with specified cover and management to that from an identical area in tilled continuous fallow

    P, the support practice factor, is the ratio of soil loss with a support practice like contouring, strip-cropping, or terracing to that with straight-row farming up and down the slope

Numerical values for each of the six factors (*RKLSCP*) have been determined from analyses of the soil loss and runoff data and weather records. Values of the factors for all soils, sites, and management conditions in the United States have been determined and are in the process of continual refinement—and are available through the local district (county) offices of the Soil Conservation Service.

The *R* (rainfall and runoff) factor is a measure of the erosive potential of the rainfall within a locality to move the soil under long-term average climatic conditions. The *K* values are dependent upon the characteristics of the soil profile (Table 40). Soils high in silt are generally most erosive, and soils least erosive are high in sand or clay. Some soils with high content of very fine sand are also very erosive. Texture, structure, organic matter content, and permeability are important soil profile characteristics that interact together to determine how easily or severely a soil will erode.

TABLE 40 *Computed* K *values for specific soils on some of the erosion research stations (adapted from Wischmeier and Smith, 1978)*

Soil and condition	Source of data	Computed K value
Dunkirk silt loam (fallow)	Geneva, NY	0.69
Keene silt loam (rowcrop)	Zanesville, OH	0.48
Shelby loam (rowcrop)	Bethany, MO	0.41
Lodi loam (rowcrop)	Blacksburg, VA	0.39
Fayette silt loam (fallow)	LaCrosse, WI	0.38
Cecil sandy clay loam (rowcrop)	Watkinsville, GA	0.36
Marshall silt loam (rowcrop)	Clarinda, IA	0.33
Ida silt loam (rowcrop)	Castana, IA	0.33
Mansic clay loam (rowcrop)	Hays, KS	0.32
Hagerstown silty clay loam (fallow)	State College, PA	0.31
Austin clay (rowcrop)	Temple, TX	0.29
Mexico silt loam (rowcrop)	McCredie, MO	0.28
Honeoye silt loam (fallow)	Marcellus, NY	0.28
Cecil sandy loam (fallow)	Clemson, SC	0.28
Ontario loam (fallow)	Geneva, NY	0.27
Cecil clay loam (rowcrop)	Watkinsville, GA	0.26
Boswell fine sandy loam (rowcrop)	Tyler, TX	0.25
Zaneis fine sandy loam (rowcrop)	Guthrie, OK	0.22
Titton loamy sand (rowcrop)	Tifton, GA	0.10
Freehold loamy sand (rowcrop)	Marlboro, NJ	0.08
Bath channery silt loam (fallow)	Arnot, NY	0.05
Albia gravelly loam (rowcrop)	Beemerville, NJ	0.03

Figure 50 illustrates how soil profile characteristics influence erosion. From Figure 50 one can estimate K values for soils with assumptions that the soil has about 2 percent organic matter, fine granular structure, moderate permeability, and very fine sand comprising about 20 percent of the sand fraction. Relative modifications and adjustments in K values can be made for individual soil profile conditions unique to each soil map unit (Ahn, 1978). The most erosive soils have K values in the higher ranges; the lines separating the more clayey textures and the more sandy textures segregate out the remaining textures that are most erosive (silt, silt loam, loam; see Fig. 11). The K values for soils are increasingly being published in soil survey reports, so that these values are becoming more readily available. If the K values are unpublished, they can be obtained for the local soils in a survey area from the District Office of the Soil Conservation Service.

Table 41 shows some relative K values of some soils in Hawaii, according to categories of Soil Taxonomy (Soil Survey Staff, 1975). Each soil series name represents unique characteristics of each soil profile which is erosive or not so erosive. The Family groups have particle-size and clay mineralogy designations that help to segregate erosion classes of soils (Fig. 14). Higher categories (Subgroup, Great group, Suborder, Order) are broader, and give less information that can be related to specific K values or erosion classes.

Slope length and steepness (LS) is recorded by soil map unit areas on aerial photographs, and can be interpreted from soil maps for farm planning purposes to help control erosion. Table 42 gives some values for slope lengths and steepness. Steeper slopes have greater soil erosion potential because runoff is more rapid, and longer slopes have more water-gathering capacity to move larger volumes of sediment. Slopes can be made shorter and less steep with terraces.

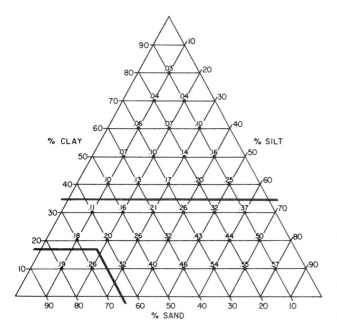

FIGURE 50 *Soil texture triangle with* K *values plotted on it to show approximate relationships between sizes of mineral particles and soil erodibility (adapted from Ahn, 1978).*

The cover and management factor (C) and the support practice factor (P) are the determinants in the soil-loss equation by which erosion can be controlled. Basically, soil is protected from erosion by vegetation, which reduces the kinetic energy of falling raindrops by intercepting their impact upon the soil, which would otherwise detach and erode the soil particles. Forest or grass cover protects the soil surface throughout the growing season so that erosion is minimal. Row crops and small grains leave part of the soil surface bare for certain periods, so that the soil is less well protected. Fallow leaves the soil unprotected by growing vegetation, but mulch at the surface will partly protect the soil. Combinations of contour grass strips, grassed waterways, pasture, and other protective crops will help to reduce soil erosion. Often, row crops can be put on the nearly level soils that do not erode so readily, and pasture, trees, and grass or hay crops can be put on the steep soil slopes to protect the more damage-susceptible soils from excessive erosion. Selecting the best combinations of crops, soils, and slopes is the essence of farm planning, together with terraces, diversions, and waterways to control excess water that must flow from the land during prolonged heavy rains.

Soil erosion is a natural geomorphic process, of course, so that it can never be eliminated completely. For each soil, tolerable soil losses have been established that will permit sustained agricultural production over a long period of time. Deep sandy soils with uniform profiles can tolerate up to 5 tons of soil loss per acre per year without serious yield reductions. Soils shallow over bedrock or hardpan, in contrast, cannot have more than 2 tons of soil loss per acre per year without serious lasting damage. With dams, ponds, grassed waterways, and other conservation storage features, most sediment loss can be trapped close to the source of the erosion. In this fashion soil loss can be minimized, sediment can be controlled, and flooding downslope on other farms and in urban areas can be greatly reduced. Soil erosion and runoff control are in the best interests of everyone. Soil-loss-equation calculations have recently been adapted to urban construction areas where erosion is severe, especially when the soil surface is bare before vegetation can be established.

TABLE 41 *Approximate K values for 10 benchmark soils in Hawaii according to categories of Soil Taxonomy (adapted from Foster, 1977)*

Order	Suborder	Great Group	Subgroup	Family	Series	K value
Ultisol	Humult	Tropohumult	Humoxic Tropohumult	Clayey, koalinitic, isohyperthermic	Waikane	0.10
Oxisol	Torrox	Torrox	Typic Torrox	Clayey, koalinitic, isohyperthermic	Molokai	0.24
Oxisol	Ustox	Eutrustox	Tropeptic Eutrustox	Clayey, koalinitic, isohyperthermic	Wahiawa	0.17
Vertisol	Ustert	Chromustert	Typic Chromustert	Very fine, montmorillonitic, isohyperthermic	Lualualei	0.28
Aridisol	Orthid	Camborthid	Ustollic Camborthid	Medial, isohyperthermic	Kawaihae	0.32
Inceptisol	Andept	Dystrandept	Hydric Dystrandept	Thixotropic, isohyperthermic	Kukaiau	0.17
Inceptisol	Andept	Eutrandept	Typic Eutrandept	Medial, isohyperthermic	Naolehu	0.20
Inceptisol	Andept	Eutrandept	Entic Eutrandept	Medial, isohyperthermic	Pakini	0.49
Inceptisol	Andept	Hydrandept	Typic Hydrandept	Thixotropic, isohyperthermic	Hilo	0.10
Inceptisol	Tropept	Ustropept	Vertic Ustropept	Very fine, kaolinitic, isohyperthermic	Waipahu	0.20

TABLE 42 *Values of the topographic factor (LS) of the soil-loss equation for specific combinations of slope length and steepness (adapted from Wischmeier and Smith, 1978)*

Slope (%)	Slope length (ft)				
	50	100	200	500	1,000
2	0.16	0.20	0.25	0.33	0.40
4	0.30	0.40	0.53	0.76	1.01
6	0.48	0.67	0.95	1.50	2.13
8	0.70	0.99	1.41	2.22	3.14
10	0.97	1.37	1.94	3.06	4.33
12	1.28	1.80	2.55	4.04	5.71
14	1.62	2.30	3.25	5.13	7.26
16	2.01	2.84	4.01	6.35	8.98
18	2.43	3.43	3.86	7.68	10.90
20	2.88	4.08	5.77	9.12	12.90

In making farm plans to control erosion, the soil map is studied carefully by a soil conservationist. Soil map units are grouped and colored according to the land capability classes. The conservationist talks to the farmers about their ideas for production goals from their fields and farm. By mutual agreement, soils and crop systems are matched with engineering structures (dams, terraces, waterways) for a program of soil and water conservation on each farm. The economic implications are critical, and expensive erosion control measures must be established on a cost-sharing basis for the more expensive inputs. Technical assistance is provided to the farmer where necessary. Throughout the process, educational programs are in progress to educate urban populations as well as rural people about the importance of soil conservation to the society. In this manner, through cooperative and concentrated efforts, soil maps are established as the base for the inventory of our most valuable resources. Integrated social programs are being designed to integrate that knowledge for most efficient land management.

EVIDENCE OF SOIL EROSION

A great deal of evidence is available to show that excessive soil erosion is detrimental to everyone. Figure 51 illustrates a buried soil profile that was farmed by the Maya in the Valle de Naco in Honduras more than 1,000 years ago. The Maya had a dense population that expanded into the hills around the valley site shown in Figure 51. As the population expanded, forests were cleared in the uplands and soil erosion was greatly accelerated. Eventually, excessive erosion caused flooding in the valley, and yields declined in the uplands. About 1,000 years ago the civilization collapsed, and the population abruptly declined. The alluvial streambank cut in Figure 51 records some of the events of the past Maya history. The lower profile which the Maya used has artifacts of the period scattered through the soil and accumulations of phosphorus from the human habitations; the upper profile is sterile of artifacts. Apparently, the erosion depleted the soil resource base and caused flooding, so that production declined and could no longer support the dense Maya population. When the population declined, the vegetation recovered—and the cycle appears to be repeating with the modern land abuse. The present population in the area is probably less than in the Maya period, but it is increasing. This history, recorded in

FIGURE 51   *Buried soil profile in alluvium used by the Maya in Valle de Naco in Honduras more than 1,000 years ago and abandoned when excessive flooding inundated the valley. Sediments about 1 m thick were deposited over the Maya soil, in which a modern soil has formed. The destructive flooding and sediment was caused by excessive clearing of trees and soil erosion in the uplands.*

the soils, is a message to be heeded by modern planners and soil conservationists for the future.

CANADARAGO COMPUTER STUDY

With computers and modern soil maps, soil erosion can be monitored and controlled much more closely in the future than it has been in the past. In an example study (Kling and Olson, 1975), the drainage basin of Canadarago Lake in east-central New York State was divided into 4,193 ten-acre cells for study of soil erosion movements (Fig. 52). For each cell, entries were made on a computer card to describe for that cell the erodible characteristics of the representative soil. Factors of the soil-loss equation were determined from the soil map, land-use map, and topographic map. A computer program was developed to estimate the soil erosion toward the streams and lake. The erosion and deposition were estimated by computer for each cell, and each cell was given designations to indicate where it should have received sediments and where it should have passed on erosion materials to the next cell. The group of cells shown in Figure 52 is one of the 771 ephemeral drainage nets delineated for the drainage basin for the computer.

A mathematical transport factor was used to estimate the efficiency of transport of eroded material between adjacent cells within the drainage networks. Through the use of this transport factor and the drainage networks, it was possible to calculate an estimate of the gross erosion for each cell (determined from the soil-loss equation), the net erosion of each cell (the estimated amount of material actually moving out of each cell), and the

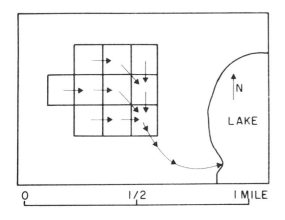

FIGURE 52 *Directions of soil movement from cell to cell to stream to lake in a part of the drainage basin of Canadarago Lake (adapted from Kling and Olson, 1975).*

net yield of each cell (the estimated amount of material that finally reaches the streams or lake from each cell). Estimates and computer calculations were checked with monitorings and analyses made from the four major streams emptying into the lake.

Table 43 is a summary of the computer calculations for the drainage basin, indicating estimated sediment losses to the lake as related to land uses in the basin. The "erosion factor" for each land use in Table 43 is the factor of the soil-loss equation determined by relative erosion hazard of each land use. The erosion factor is closely related to the cover or vegetative material protecting the soil particles from detachment. Relative erosion potentials are: croplands (0.300) > inactive croplands (0.080) > brushlands (0.050) > pastures (0.030) > forest lands (0.020) > pine plantations (0.004) > wooded wetlands and bogs (0.000). Gravel pits and stone quarries (1.000), although small in area, have bare

TABLE 43 *Estimated sediment yield to streams and lake calculated by computer for one year from selected land-use areas in Canadarago Lake drainage basin (adapted from Kling and Olson, 1975)*

Land use	Soil erosion (tons)	Sediment (% of total)	Erosion factor	Land-use area (% of total)
Croplands	107,380.3	78.4	0.300	34.6
Forest lands	6,644.7	4.9	0.020	16.6
Brushlands	13,877.1	10.1	0.050	16.4
Pastures	6,292.0	4.6	0.030	15.9
Inactive croplands	2,138.6	1.6	0.080	4.9
Wooded wetlands	0.0	0.0	0.000	5.0
Bogs	0.0	0.0	0.000	2.5
Pine plantations	86.3	0.1	0.004	1.1
Shoreline developments	93.9	0.1	0.030	0.8
Low-density residential	56.3	< 0.1	0.006	0.8
Rural hamlets	26.4	< 0.1	0.006	0.5
High-density residential	17.2	< 0.1	0.006	0.2
Gravel pits	203.2	0.1	1.000	0.1
Stone quarries	80.2	0.1	1.000	0.1
Railways	14.3	0.1	0.150	0.1
Medium-density residential	6.6	< 0.1	0.006	< 0.1
Residential strips	2.7	< 0.1	0.030	< 0.1
Central business sections	2.6	< 0.1	0.006	< 0.1

surfaces exposed throughout the year so that large amounts of sediment can potentially be released from them if conservation measures are not installed.

Computer programs also enable study of alternative means by which soil sediment eroded to Canadarago Lake could be reduced. The sediment transport model in the computer provided a method for theoretically evaluating the effects of proposed land-use changes on soil losses. Because current land use is represented internally within the computer, it is relatively easy to simulate land-use change by changing the land-use patterns within the computer. Almost any land-use change can be simulated by the computer at a very modest cost before going to the field with operational conservation programs.

Proposed land-use changes were simulated in the computer. Zoning a 660-ft-wide greenbelt of permanent pasture on both sides of all streams resulted in a predicted reduction of 15 percent in total sediment yield. Contour strip cropping on all cropland would reduce sediment yields about 37 percent. More expensive conservation practices, such as terracing, would reduce soil loss and sediment yields even further.

Estimated erosion in the Canadarago Lake drainage basin.

FIGURE 53 *Estimated gross erosion for each 10-acre cell in Canadarago Lake drainage basin (adapted from Kling and Olson, 1975).*

To summarize the computer calculations for the entire Canadarago Lake drainage basin, Figure 53 gives the classes of estimated erosion for each 10-acre cell, based on the soil-loss equation. Figure 53 maps soil loss only, without figuring redeposition of sediment. Areas in Figure 53 with highest soil losses are having the productive capacity of the soil most damaged by the high erosion rates. Figure 54, on the other hand, locates the sources of sediment yield ultimately reaching the streams and lake. The areas in Figure 54 that lose more than 50 tons of soil per acre per year are those most in need of immediate attention to improve the quality of the lake. Those that lose 10–50 tons are of second priority. When conservation programs are instituted on a larger scale in the basin, all the areas with greatest erosion located both in Figures 53 and 54 should receive long-term investments for conservation practices. Incentive payments and technical assistance to farmers would be a necessary part of a most effective program to improve the total environment.

Uses of soil maps in landscapes are well illustrated by erosion control procedures.

Estimated sources of sediment yield in the Canadarago Lake drainage basin.

FIGURE 54 *Estimated erosion yield from each 10-acre cell in drainage basin to streams entering Candarago Lake (adapted from Kling and Olson, 1975).*

Soil maps identify and locate the different soils of the resource base, and identify their erosion potentials. Predictions are made about future soil losses under different land uses and cropping systems. Computers are a useful tool to handle large amounts of data and make predictions. Farm planning uses soil maps in a tailored manner to fit the individual needs and desires of every farmer. Soil map units, fields, farms, and the total environment or drainage basin can be considered in the planning processes. In an integrated fashion, resources of the land can be used in a better manner in the future than they have been in the past. The key to better land use is in better use of soil resources through improved application of soil map unit information.

CHAPTER 9

YIELD
CORRELATIONS

Yield and crop performance are closely correlated with soil conditions. All the soil factors described in a soil survey and those quantified in the laboratory are vital for predicting crop yields. Figure 55 illustrates crop differences due to soil conditions. The soil in the central part of Figure 55 occupies the low, wet part of the field and has prominent mottles near the surface; textures of this soil are more clayey than in soils on the upper slopes, and water tables are near the surface in the early spring. Obviously, yields from the wet soil will be extremely low in wet years, but yields on the same soil may be higher in drier years. Measurements and observations of crop yields on individual soils over many years are used to make estimates for predictive purposes for each soil map unit. Of course, wet spots such as that shown in Figure 55 can be drained, and then the soil and the field is greatly improved in its productive potential for the future.

FIGURE 55 *Effects of soil drainage differences on corn growth in a field in northwestern Onondaga County, New York.*

119

TABLE 44 Estimated average yields per acre of some principal crops under two levels of management in the Virgin Islands (adapted from Rivera et al., 1970)[a]

Soil map unit	Tomatoes (lb)		Sweet potatoes (lb)		Grain sorghum (lb)		Silage sorghum (tons)		Guineagrass (tons)	
	A	B	A	B	A	B	A	B	A	B
Coamo clay loam, 2–5% slopes	7,000	8,500	7,000	9,000	4,000	5,000	19	22	9	11
Cornhill gravelly clay loam, 0–2% slopes	5,500	7,000	6,500	8,500	3,500	4,500	18	21	8	10
Fredensborg clay, 0–2% slopes	7,500	9,000	8,000	10,000	5,000	6,000	22	25	12	14
Fredensborg clay, 2–5% slopes	7,000	8,500	7,000	9,000	4,300	5,300	20	23	10	12
Fredensborg clay, 5–12% slopes, eroded	6,500	8,000	6,000	8,000	4,000	5,000	15	18	7	9
Glynn clay loam, 2–15% slopes	8,000	9,500	8,000	10,000	5,000	6,000	22	25	12	14
Glynn clay loam, 5–12% slopes, eroded	6,500	8,000	7,000	9,000	4,300	5,300	20	23	9	11
Hesselberg clay, 0–2% slopes	6,500	8,000	6,500	8,500	4,000	5,000	17	20	9	11
San Anton clay loam, 0–3% slopes	8,000	9,500	8,000	10,000	5,000	6,000	22	25	12	14
San Anton clay loam, 5–12% slopes	6,500	8,000	7,000	9,000	4,300	5,300	20	23	9	11
Sion clay loam, 0–5% slopes	6,500	8,000	6,500	8,500	4,000	5,000	17	20	9	11

[a]Column A indicates yields under average management; column B indicates yields under a high level of management.

TABLE 45 *Suitability of fruit trees, shade trees, exotics, and ornamentals for specified soils on Saint Croix (adapted from Rivera et al., 1970)[a]*

Plants	Coamo cl	Cornhill gr cl	Fredensborg c	Glynn cl	Hessleberg c	San Anton cl	Sion cl
African tulip	2	2	2	2	3	1	2
Alamanda	1	1	2	2	3	1	2
Almond	2	2	2	2	3	1	2
Australian pine	1	1	1	1	2	2	1
Avocado	2	2	1	3	3	1	1
Bamboo	2	2	2	2	3	1	1
Banana, plantain	1	1	1	2	3	1	1
Bougainvillea	1	1	1	2	3	1	1
Breadfruit	3	3	3	3	3	2	2
Cashew	1	3	2	3	3	1	2
Cedar (tabebuia)	2	2	2	2	2	1	1
Coconut palm	2	2	2	2	2	1	2
Cocoplum	1	3	2	2	3	1	2
Croton	3	1	2	3	3	1	1
Custardapple	3	3	3	3	3	1	2
Date	2	2	2	2	2	1	2
Flamboyanttree	1	1	2	1	2	1	1
Fragipani	1	1	1	1	2	1	1
Genip	1	2	1	1	2	1	1
Ginger Thomas	1	1	1	2	2	1	1
Gooseberry	1	2	1	1	2	1	1
Guava	1	2	2	2	3	2	2
Guavaberrytree	2	2	3	1	2	1	2
Hibiscus	2	2	1	2	2	1	2
Hogplum	1	2	2	1	2	2	1
Jerusalemthorn	1	1	1	1	2	1	1
Jujube	1	1	1	1	2	1	1
Lime	1	1	1	2	2	1	1
Mahogany	1	1	1	1	2	1	1
Mamey	1	1	1	1	3	1	1
Mango	1	2	2	2	3	1	3
Mesple	1	2	1	2	3	1	1
Oleander	1	2	1	1	2	1	1
Orange, grapefruit	2	2	2	2	3	3	2
Papaya	1	2	1	2	3	1	1
Pineapple	1	3	2	2	3	2	2
Pomegranate	2	2	2	2	3	1	2

[a] 1 indicates that the plant is suited to the soil, 2 indicates that it is suited under special management, and 3 indicates that the plant is not suited to that soil.

ESTIMATED YIELDS

Estimated yields are commonly published in soil survey reports for all crops grown in a survey area. Usually, the estimates are made from data of agricultural experiment stations, and measurements and observations made by cooperative extension agents, soil conservationists, soil surveyors, and others. To an increasing extent, however, yields are being measured on farmers' fields under actual conditions of weather and management over long periods of time on a statistical sampling basis.

Table 44 is an example of estimated yields for tomatoes, sweet potatoes, sorghum, and guineagrass on some soil map units in the Virgin Islands. Tomatoes commonly yield only 5,500 lb/acre under average management on Cornhill gravelly clay loam on 0-2 percent slopes, but can yield 9,500 lb per acre on San Anton clay loam on 0-3 percent slopes under a high level of management. Sweet potato yields range from 6,000 to 10,000 lb per acre on different soils, and guineagrass yields range from 7 to 14 tons per acre. These data are important for planning about imports and exports, and for planning operations of even very small farms. With these data a farmer can predict probable yields in an average year, and set goals toward yields that could be achieved through improved management.

When yield data are not available or not needed (as for ornamental trees), then general interpretations can be made about plant performance on different soils as illustrated in Table 45. Fragipani trees are well suited to Coamo clay loam, but suited only under special management for Hesselberg clay. Orange and grapefruit trees are not suited for Hesselberg clay, but are suited to Coamo clay loam under special management. Coamo clay loam is good for growing papaya and pineapple, but poor for breadfruit.

SOIL CORRELATIONS

Soil scientists are always searching for yield correlations to specific soils and soil properties, and for correlations of soil characteristics with each other. Grouping and classification of soils provides a convenient technique to use to make samplings in broad soil areas. Sarkar et al. (1966) used correlation coefficients to successively eliminate soil characteristics interrelated with others—and to select the most independent soil properties. From data on 26 soils of nine Orders, 61 soil characteristics were selected. From the list (Sarkar et al., 1966), the 22 most independent soil characteristics were:

1. Structure of B2
2. Thickness of A1
3. Thickness of B2
10. Degree of mottling
11. Fe-Mn concretions
12. Depth to rock or permafrost
13. Thickness of organic layer above A horizon
15. Average percent slope
16. B2 consistency
20. Chroma of A
21. Hue of B
22. Value of B
23. Chroma of B

30. Percent clay in B2
31. Percent clay in A1/percent clay in B2
33. Percent organic carbon in A1
34. Percent organic carbon in B2
38. pH of B
42. Cation-exchange capacity of A/cation-exchange capacity of B
48. Extractable Na in B/extractable Na in C
52. $\dfrac{\text{Very fine sand} + \text{coarse silt of B}}{\text{Fine silt}}$
53. $\dfrac{\text{Coarse silt of B}}{\text{Fine silt}}$

These 22 soil characteristics are applicable only to the 26 soils under study, but they are probably representative of most of the more independent soil properties because the 26 soils represent a broad range. Many of the soil characteristics are related to moisture-holding capacity, organic matter, pH, and cation-exchange capacity, which are critical to plant growth and crop production.

Soil test results are closely related to soil map units, and the correlations of the field and laboratory data can produce extremely useful generalizations for predictions of future yields. Westin (1976), in South Dakota, determined correlations for more than 80,000 farmer samples with soils classified at the Series, Family, Subgroup, Great Group, Suborder, and Order categories of Soil Taxonomy. With higher carbonates in soils, potassium and phosphorus decreased and pH increased. Soils on steeper slopes had lower levels of potassium, phosphorus, and organic matter, but higher pH values. Soils in western Montana, measured for rangeland yields (Munn et al., 1978), fit into the sequence shown in Figure 56. From 23 sites it appeared that thickness of the A1 horizon (Mollic epipedon) was the soil characteristic that correlated most closely with productivity.

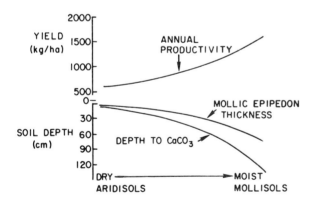

FIGURE 56 *Relationships between Mollic epipedon thickness and depth to calcium carbonate in soils of western Montana and resultant rangeland yields (adapted from Munn et al., 1978).*

Increasingly, soil scientists are being requested to work with crop scientists, agronomists, foresters, range specialists, ecologists, economists, and many others to measure yields being produced from different soil map units. Soil yields are being used for tax assessment, for land-use planning, for making management recommendations, and for many other purposes that demand direct linkages of production units to soils (Olson,

1979). Bell and Springer (1958) have excellently outlined a program of sampling yields on different soils, which has been very successful in Tennessee.

In Maryland, a corn yield study was conducted on 12 major soil series from 1970 to 1973 (Robinette, 1975). Sampled plots of two rows of corn each 20 ft long were randomly located and replicated four times in farmers' fields on the selected soil map units; a total of 202 sites was harvested. Actual measured yields varied somewhat from the estimated yields published in soil survey reports; in general, yields on Coastal Plain soils had been overestimated and those on Piedmont and Appalachian soils underestimated. Yields on different soils within the same field showed clear results of different soil moisture and erosion.

In Iowa, the most extensive effort to measure corn yields on soil map units was conducted in 15 counties from 1957 to 1970. Henao (1976) summarized regression modeling of data from about 2,800 plot-years. Plots were 1/100 acre in size within quarter sections randomly located. Ear corn was harvested, weighed, and a 300-g sample was taken to determine moisture content, and weight equivalents were calculated to convert ear corn to shelled corn at standard 15.5 percent moisture. Soil variables directly related to yields were slope, biosequence, available water-holding capacity, erosion, organic carbon, drainage class, percent clay, bulk density, pH, available phosphorus, and available potassium. Computer manipulations also enabled significant soil behavior (like moisture holding capacity) to be compared with soil survey data including percent sand, silt, and clay, and textural classes. Figure 57 illustrates how curves showing available water (moisture-holding capacity) of soil were plotted by the computer on the textural triangle. As in Figure 50, adjustments or assumptions need to be made about variations in soil organic matter, structure, permeability, and distribution of some of the particle-size separations.

In North Carolina, 441 plots four corn rows wide and 25 ft long were located on a grid system with about one sample plot per acre, during the period 1966–1968. Plots on a grid enabled soils to be sampled as a continuum on selected soil map units. Regression

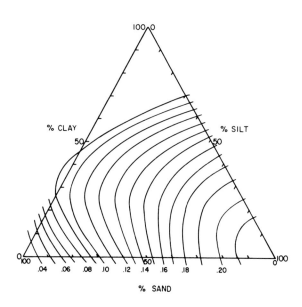

FIGURE 57 *Curves showing available water (in. H_2O/in. soil depth) in relation to sizes of mineral particles in Iowa soils (adapted from Henao, 1976). Compare this diagram with Figure 11 for the textural classes. Adjustments for organic carbon content are: 2% OC, add 0.01 in./in.; 3% OC, add 0.02 in./in.; 4% OC, add 0.03 in./in.; and 5% OC, add 0.04 in./in. For loamy sand and sandy loam, subtract 0.01 in./in. if coarse sand predominates; add 0.01 in./in. if fine sand is dominant; and add 0.05 in./in. if very fine sand is in the higher proportions.*

analysis indicated that the soil factors most highly correlated with corn yields on both Udultic and Aquultic soils were moisture-holding capacities, certain combinations of clay and sand, extractable phosphorus, percent base saturation and related properties that control soil acidity, and the amount of charge on the cation-exchange complex (Sopher and McCracken, 1973).

NATIONAL PROGRAMS

National programs exist within the Soil Conservation Service of the U.S. Department of Agriculture for relating yields to soils for woodlands and windbreaks. Production data for woodlands are related to soil map units and soil ordination class, subclass, and suitability group. Data from more than 15,000 woodland plots representing soils from nearly every state have been entered into the computer system. Effects of soils on tree growth are excellently illustrated and can be readily observed where plantations have been recently cleared across contrasting soil conditions (Buol and Davey, 1969).

Thorough discussions have been written by Shiflet (1972) of plot sizes for sampling yields of various types of natural vegetation around a soil pit. Plot sizes of 1.92, 2.40, 4.80, 9.60, and 96.0 ft^2 are well suited if vegetation is short and relatively uniform (vegetation from a plot 9.6 ft^2 weighed in grams and multiplied by 10 gives a value equivalent to pounds per acre). Larger plots must be used if trees or shrubs are included; a 66 $\times$ 66 ft plot $1/10$ acre in size is suitable. If vegetation is uneven, a more accurate sample can be obtained by using a $1/10$ acre plot 4.356 ft wide and 1,000 ft long. For statistical analysis, 10 plots $1/100$ acre in size are superior to a single $1/10$ acre plot. A national computer program is also in use in the Soil Conservation Service for relating rangeland yields to specific soils.

The Crop Reporting Service of the U.S. Department of Agriculture samples about 12,000 plots annually on major crops in 34 states, but these yields are not reported according to soil map units. It would appear that recording of soil names at each crop sampling site would be a great help in soil and yield correlation work. In the summer of 1979 a test project was underway in northern Missouri to test the feasibility of recording soil information at these sample sites. For corn one 15-ft row is harvested and weighed from each of two plots per field randomly located; two ears are sent to a state laboratory for moisture determinations. Many additional crop observations are made through the growing season.

CROP RESPONSES

Soil and crop variability are most easily observed where fertility levels are low. In Brazil (Martini, 1977), observations were made on corn plant height and yields in relation to nutrient deficiency symptoms and soil test results. Table 46 illustrates some of the data from the Oxisols and Ultisols—soils relatively low in nutrients and highly leached. Aluminum toxicity is one of the major soil factors causing yields to be low in Brazil; as aluminum increases, the yields decrease (Table 46). Calcium and magnesium are relatively low in these soils, but yields are higher where the soil test values are higher. The phosphorus test is closely correlated with yield. Correlations of soil tests with plant growth and crop yields are so good that combined tests and field observations can be used as a means to calibrate the soil test results for fertilizer recommendations. The natural

TABLE 46   *Relationships between corn plant height at flowering and yields as influenced by some soil factors in Ultisols and Oxisols in Brazil (adapted from Martini, 1977)*

Site no.	Plant height (cm)	P (ppm)	Ca + Mg (meq/100 g)	Al (meq/100 g)	Yield (kg/ha)
13	415	25.0	4.1	0.1	4,980
15	420	15.5	8.8	0.4	4,975
5	410	9.6	3.3	0.1	4,930
1	385	5.4	3.1	0.6	4,520
21	405	14.1	7.7	0.6	4,190
17	395	7.6	6.2	0.3	3,870
7	360	12.8	1.9	1.8	3,775
20	315	5.8	3.4	1.2	2,920
28	295	6.7	5.5	1.6	2,830
2	150	4.0	2.6	2.5	1,015
24	95	2.9	2.4	2.3	720
29	105	3.6	1.4	3.2	715
6	95	4.3	0.7	3.0	690
14	90	6.0	0.6	3.5	545
16	85	2.1	1.1	4.7	430

"Cerrado" or savanna vegetation is also closely correlated with the soil conditions (Lopes and Cox, 1977). Pioneer settlers in Brazil often locate the better soils by observing where the more dense stands of Cerrado trees are growing.

Weather and management are two major variables that influence the soils and the resultant yields. Torrential rainstorms, hail, drought, weeds, insects, and many other factors can destroy or damage crops very quickly. Soils, however, determine to a large extent the potential for moisture storage to withstand drought and other crop factors affecting yields. Weeds, insects, and other pests are often influenced markedly by soil characteristics. Soil factors influence runoff (Fig. 58) and erosion from storms and have a direct influence on yields. The complexity of all the soil physical and chemical interactions plus the variable weather conditions plus the many management variables make the study of yield responses to soils very difficult. With statistical sampling techniques adapted to carefully selected soils, however, it is possible to isolate the soil properties most important for determining crop yields—and to use these data in predicting future yields and potentials for improvements.

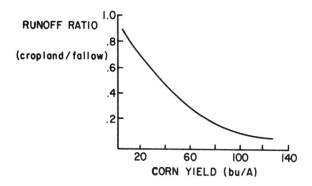

FIGURE 58 *General relationships of cornfield runoff to crop yield (adapted from Wischmeier, 1966).*

EXPERIMENT DESIGN

Unfortunately, many agronomic experiments in the past have been laid out on the assumption that the soil was uniform, or else the experimental design was structured to mask the soil differences in the statistical analysis. An example of this approach has been provided by Fly and Romine (1964) for the USDA Central Great Plains Field Station at Akron, Colorado. The 25-acre plot area studied for 46 years of uniform management included 56 dryland rotations and tillage treatments on 152 plots, 8 X 2 rods in dimensions. The field appears relatively uniform on the surface and was mapped as one soil type in a soil survey published in 1947. Mapping on a 100-ft grid at large scale on a base map with a 0.2-ft contour interval, however, showed that the area had soils that did indeed vary in color, thickness, texture, structure, depth to carbonates, and other significant properties. When the plots were indexed according to the eight soil map units finally identified, relative wheat production levels varied on those soils from 73 to 120 index units. Wheat under continuous culture on deep soils produced 20 percent more paying crops and 55 percent more wheat above cost of production than the thin soils in the experimental field plots of the same field. Soil differences induced wider variations under uniform treatment than different treatments on a single soil map unit. This study is of extreme importance, because the concept of "technology transfer" from an experiment station to other areas is dependent upon the soil identification. When the detail of delineation and characterization of soils is comparable to the detail of the research plot layout, only then can existing research fields and plot data be used to associate soil qualities with response to treatment. The reliability of application of research findings to other areas will depend on adequate characterization and evaluation of the soil in each plot. In the early days, when most experiment stations were initially sited, soil mapping was not sophisticated or sufficiently detailed to provide the kind of soil information now demanded by the need for technology transfer to large areas characterized by modern detailed soil maps.

SEQUENTIAL TESTING

"Sequential testing" is a term that has been used to refer to sampling of soils across a landscape where the soils occupy "sequential" positions in relation to their respective location and gradient of properties in the landscape. Thus soils in humid regions commonly have a "drainage sequence" in similar geologic materials (well drained, moderately well drained, somewhat poorly drained, poorly drained, very poorly drained); similar sequences can be observed over landscapes for soil texture, slope, pH, fertility, land use, crop growth, yields, and many other soil and land characteristics. Sequential testing of soils consists of studying the soils and their characteristics (including yields) according to the relationships each soil has with the others. Thus, red pine (sensitive to soil wetness) will exhibit growth according to soil drainage classes, and sequential testing can measure the tree growth in relation to the soil depth to mottling and depth to seasonal water tables. Sequential crop sampling is particularly valuable because yield differences can be sampled that are due strictly to the soils. Soils in single fields where weather and management are constants are the only variables in the sampling "experiment." Thus, yield effects strictly due to the soils can be isolated by sequential testing.

Sequential testing is essentially a reversal of the idea of locating experiments or data collection points on uniform soils; sequential testing consists of deliberately locating experiments and data collection points on contrasting soils to evaluate selectively the effects of soil differences. In sequential testing, the soils are considered to be the basic resources, and the cropping and teatment design is fit into the soil characteristics of the landscape. Experiments or data sampling points can be established along a transect (or a grid) deliberately traversing soil boundaries. The experiment may be on a macro or micro scale to measure or sample land use, texture, drainage, slope, pH, fertility, or other characteristics of the soils. The soils isolated and studied depend upon the nature of the landscapes and the soils that occupy their relative parts of landscapes. Along a sampling transect or "strip," complex experiments with monitoring points can be installed, or the sampling can be reduced to its simplest form by only "paired" or "triplicate" samplings of points within a single field.

Examples of sequential sampling of soils and yields can be found in the literature. Malo and Worcester (1975) described soil and yield studies along a transect from a topographic summit into a pothole in glacial till in the Red River Valley. A diagram of their transect and sampling points is given in Figure 59. One-foot contour intervals are

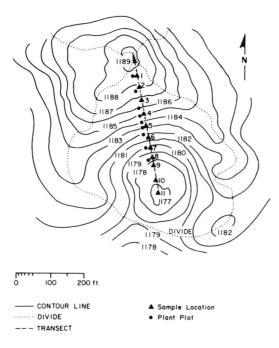

FIGURE 59 *Soil and crop yield sampling transect from a topographic summit to a North Dakota landscape depression (adapted from Malo and Worcester, 1975).*

marked on the map, and samplings were made of barley and sunflower yields. The variable soils along the transect included Barnes (Summit Position—Udic Haploboroll), Buse (Shoulder Position—Udorthentic Haploboroll), Hamerly (Footslope Position—Aeric Calciaquoll), Svea (Backslope Position—Cumulic Haploboroll), and Tonka (Toeslope Position—Argiaquic Argialboll). Sunflower yields across the transect are plotted in Figure 60. Close relationships existed between the soil properties and the plant responses in the various landscape positions. These soils and related soils are repeated in sequences across large areas of these types of regional landscapes, so that the data of yield variabilities can be extended over large areas. In fields managed similarly, the different soils give very different responses, and these responses are reflected in differences in quality and

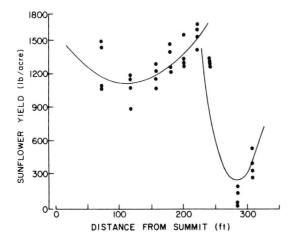

FIGURE 60 *Sunflower yields from sequential testing in a transect from a topographic summit to a North Dakota landscape depression (adapted from Malo and Worcester, 1975).*

quantity of yields. Different sequential management could produce a more uniform crop in these fields.

An excellent example of sequential testing in the ultimate sense is provided by Wallen and Jackson (1978). A plot measuring 197.5 × 43.9 m was planted to 10,633 alfalfa plants (49 plants per row spaced 0.9 m apart, in 217 rows). The plants were then evaluated periodically for their survival during frost heaving over several seasons, on four soil map units which were delineated at large scale within the plant observation plot. From an altitude of 105 m aerial photographs on 70-mm film were taken from a helicopter, to record the seasonal conditions of each plant. Over several seasons the monitored plants clearly showed the effects of the soil conditions; well-drained soils supported a good population of plants, but poorly drained and moderately well drained soils had high plant mortality. This method of detailed aerial surveillance even disclosed some spots within the well-drained soil map units which had restricted water movement which in turn affected the alfalfa survival. This type of detailed sequential testing of soils is a very useful method which can provide keen insights into the behavior of plants. The study is an example of the use of initiative and inventiveness in the sequential testing of soils.

In farmers' fields, yields are being measured for different crops on selected soil map units in programs in several states (Illinois, Iowa, Minnesota, Missouri, Maryland, Tennessee). Paired samplings of yields on highly contrasting soils in a single field permit isolation of weather and management variables. In this kind of soil research, the world is regarded as a gigantic research station, and the experiment consists of carefully selecting the sites with the constraints and variables to be studied. For yield correlations with soils, it is most important to identify and map the soil differences completely, and then measure the crop responses to these soil characteristics. In this fashion, yield data can be extended (through "technology transfer") to other areas of the same or similar soils with similar climate and management. Predictive implications of the extension of these data are enormous. Many of the agricultural improvements in regions are dependent upon yield correlations to soil map units.

CHAPTER 10

ARCHEOLOGICAL CONSIDERATIONS

The past is a key to the future. In archeological strata there are artifacts that record events and processes during millennia of human occupations of different soils. In many areas, contemporary use of soils is not greatly different from ancient use of soils; in other places technology has changed a great deal, but always the soils in landscapes have a dominant influence upon the land-use patterns. Often, the archeological record proves that ancient people abused their soils, and that their civilizations were disrupted by the ecological and environmental consequences. In places, the ancient populations were larger than the modern populations; the archeological considerations give reasons for us to ponder if we ourselves are using our resources most wisely, and to seek explanations for the ancient population declines and shifts so that we might avoid a similar fate. Generally, the major and most important factors in the continuing prosperity of a nation are not its conquests in war, but the nature of the husbandry of the soil and land resources through centuries of use and occupation.

NEW YORK

In New York State the earliest occupants of the land were Paleo-Indian hunters; a type archeological site for the hunting stage was located in well-drained soils in glacial till on a drumlin surrounded by swampy soils which were probably a shallow lake about 10,000 years ago. From about 3500 to 1300 B.C. people began to fish and gather wild vegetable food as well as hunt, and then they selected settlement sites on well drained soils on beach ridges and outwash terraces. From 1300 to 1000 B.C. Indians began to use soil and rock materials to make pottery and simple tools, in addition to arrowheads and scrapers. Soils came into use for agriculture and village life from 1000 B.C. to A.D. 1600. Cemeteries were located in well-drained sandy and gravelly soils prominent in the local landscapes. Remains of a stockaded Indian village of at least 10 houses were found west of Syracuse on a low ridge of well-drained Palmyra gravelly silt loam soil, near wet soils formed by seepage from a spring. These people raised corn, beans, and squash, gathered wild plants, hunted, and fished. From evidence of abandonment, archeologists have speculated that exhaustion of local supplies of firewood, game, and soil fertility resulted in the abandonment of the village for a time until adequate natural restitution around the site had been accomplished.

MESA VERDE

In southwestern Colorado, Mesa Verde is a dissected tableland with soils that once supported thousands of people more than 1,000 years ago. The destruction of the forest and natural vegetation by the Anasazi Indians promoted soil erosion and environmental degradation, which eventually contributed significantly to the abandonment of the area. As the ancient population increased, even dams, terraces, and other water management systems could not reduce the soil loss to acceptable levels to achieve and sustain the essential agricultural production. About A.D. 1200 people began moving away, and by A.D. 1300 nearly everyone was gone. The environmental deterioration and drought, together with excessive population and social stress, is commonly blamed for the departure. A few unburied skeletons found later by early white explorers might have been the remains of a few old or sick people who refused to migrate with the others. After abandonment, the vegetation gradually recovered on Mesa Verde, and the arid climate preserved many of the tools and artifacts of the Anasazi in the soils. The lesson to be learned is that the Anasazi were not greatly unlike some of us in their application of principles of soil and water management. Ambler (1977) summarized the Anasazi outlook of the past as it relates to our own concepts in planning for the future:

> They denuded large areas of their forest cover, littered the ground in front of their houses with trash, caused tons of topsoil to be eroded away and depleted of nutrients, and filled their lungs with soot from the fires. When things got too tough or used up in one spot, they moved to another and probably sometimes moved just because they became bored with the same place or wanted a different view. In short, the difference between the Anasazi of 500 to 1,000 years ago and the present people of the Southwest is more one of degree than of kind.

PHOENIX

The Phoenix, Arizona, area had extensive irrigation agriculture by the Hohokam Indians in the Salt and Gila River valleys about A.D. 500. Several hundred miles of canals were constructed and have been mapped on aerial photographs. The Hohokam grew maize, jackbeans, lima beans, kidney beans, tepary, amaranth, squash, gourds, cotton, and tobacco. Eventually, the difficulties of soil management became a major problem, and soils became waterlogged because of excessive irrigation and inadequate drainage. Sediment filled the canals, and salts accumulated in the fields. Salt erosion was also a problem in the stability of the adobe walls of the villages. Salt weathering of the adobe walls made them susceptible to deterioration, thus causing their crumbling and collapse. By A.D. 1400 the irrigated areas had been largely abandoned. Many of the modern canals follow the route of the ancient Hohokam canals, but modern technology has enabled better crop production through improved water management and soil drainage. Soil erosion and salinity problems, however, remain a constant threat to the stability of even the modern irrigation systems.

TIKAL

Extensive research on relationships of soils to archaeology has been done in Maya areas. At Tikal in the Peten region of Guatemala, soils were described, sampled, and

FIGURE 61   *Vertisol soils in bajo swamps at Tikal. These soils mapped with the 10/A symbols are extremely sticky and expansive when wet, and shrink and crack when dry. The Maya strictly avoided these areas as construction sites for large buildings. This modern road will have many problems with trafficability, and the soils will not provide a very satisfactory roadbed in the rainy season.*

FIGURE 62   *Ruins of a Maya structure at Tikal. Heavy buildings were mostly constructed on Mollisols shallow to limestone bedrock. The city was abandoned about 1,000 years ago, and Inceptisols and rain forest now cover the ruins of the former temples, palaces, plazas, causeways, and other constructions.*

mapped at a scale of 1:2,000 (Olson, 1977). The Maya apparently achieved good mastery of the visible soil properties related to engineering use and management through experience more than 2,000 years ago; the invisible and less visible degenerative soil processes (soil fertility depletion and erosion), however, caused many insidious long-run problems for the Maya. The soil survey of Tikal is a good example to look at in detail and to generalize from because it is the most detailed (large-scale) and comprehensive soil survey ever conducted specifically for soil and archeological correlations in the Maya areas.

At Tikal, Mollisols, Vertisols, Inceptisols, Entisols, and various other soils were discovered, described, and mapped in the central core of the abandoned Maya city. The bajo swamps had Vertisols (Fig. 61) with soil properties extremely poor for heavy urban constructions; the Maya avoided these areas for their buildings. In contrast, the Mollisols shallow to limestone bedrock afforded excellent sites for support of heavy limestone buildings (Fig. 62). Figure 63 illustrates the soil pattern (at 1:2,000 scale, reduced) of the North Zone quadrangle of the abandoned Tikal city. No ruins of buildings are located in the 10/A Vertisol areas, and dense clusters of ruins are congregated on the Mollisols. The aquada water hole (11/A) located in the center of Figure 63 was a Maya water reservoir, and the sediments in the bottom record the periods of soil erosion from the Maya abuse of the surrounding higher ground.

Mollisols of the uplands at Tikal were naturally very fertile but extremely vulnerable to erosion and damage by Maya populations because of the shallowness of the soil to limestone bedrock. The dark Mollisol appearance of the thin surface soil formation misled the Maya settlers, because the topsoil was quickly oxidized and depleted by erosion when the forest was cleared. As the Maya population expanded, erosion and soil depletions accelerated until the "tragedy of the commons" forced excessive deterioration of the soil resources. In areas disturbed by the Maya including causeways and built-up areas, light-colored Inceptisols and Entisols were mapped that were distinctly different from the dark-colored undisturbed Mollisols. Areas in the uplands of Tikal disturbed by the Maya could be located and mapped in the silt loam places based on soil color alone. Light soils were depleted; dark soils were nutrient-rich. Soils at Tikal have not recovered from the Maya occupation even after more than 1,000 years of abandonment to the rain forest.

The bajo lowland Vertisols at Tikal were strikingly avoided by the Maya for construction sites (Fig. 63). These places are wet in the rainy seasons (Fig. 61), and the soils with montmorillonite clays are extremely sticky, even to pedestrian traffic. The Vertisols were used by the Maya for agricultural and other purposes as their population increased. Sherds (broken pieces of pottery) were commonly found in the soil survey borings made in the bajos, especially near the edges. The Vertisol soils could be identified by "slicken-sides" within the soils (as examined in pits) where structural peds rubbed against one another, verifying the "churning" action of the soils with cracking and expansion upon drying and wetting. Gilgai (mound-and-pit microrelief) also verified the instability of the soils, and facilitated the mapping of the shrinking and swelling areas of the Vertisol soils.

Soil mapping at Tikal (Olson, 1977) identified 11 different soils on six different slopes; some of the soil pattern is illustrated in Figure 63 for the North Zone quadrangle. The soils and slopes are:

Soil symbol	*Brief description*
1	Well-drained clay loam upland soil 45 cm to soft limestone
2	Well-drained silt loam disturbed alluvial soil 128 cm to soft limestone
4	Well-drained silt loam upland soil 28 cm to hard limestone

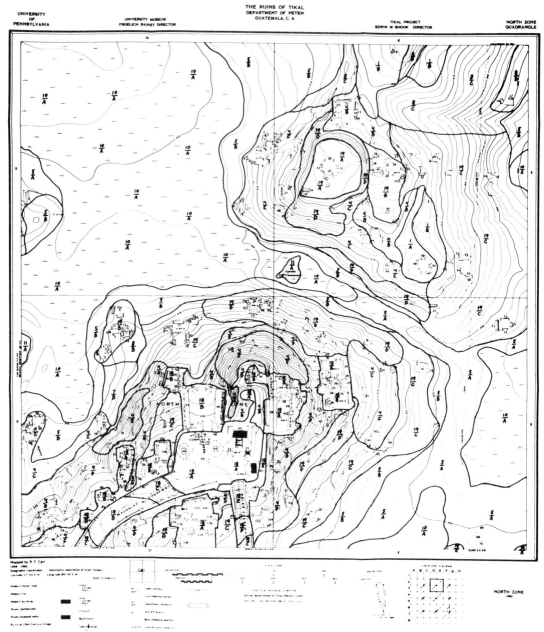

FIGURE 63 *Soil map of North Zone quadrangle at Tikal. The map has been reduced from 1:2,000 scale. The Maya strictly avoided the 10/A Vertisols as construction sites for heavy buildings, and concentrated the structures on the Mollisols shallow to limestone bedrock (adapted from Olson, 1977).*

5	Somewhat poorly drained clay swamp soil 48 cm to soft limestone
6	Well-drained silt loam disturbed upland soil 142 cm to soft limestone
7	Moderately well drained clay upland soil deeper than 150 cm to soft limestone
8	Well-drained silty clay loam upland soil 40 cm to soft limestone
10	Poorly drained clay swamp soil deep to soft limestone
11	Very poorly drained clay swamp (water hole) soil deep to soft limestone
18	Well-drained silt loam disturbed upland soil formed in soft silt loam limestone material
19	Well-drained silt loam upland soil 65 cm to moderately hard limestone

Slope designations

A	0-3% slopes
B	3-8% slopes
C	8-15% slopes
D	15-25% slopes
E	25-35% slopes
F	35-45+% slopes

Maize on the Mollisols at Tikal (in a few patches tilled at present by shifting cultivation) yields about 1,000 kg/ha in the first year after clearing, about 750 kg/ha in the second year, about 500 kg/ha in the third year, and about 250 kg/ha in the fourth year—without any fertilizer additions. For shifting cultivation, the Food and Agriculture Organization of the United Nations officially recommends a 15-year rotation on these soils—where an area cleared from the forest is cropped for 5 years, and left fallow for 10 years for forest regrowth before clearing the area again for cropping. Obviously, the Maya could not have maintained this ideal rotation as their population increased, and the consequent soil decline had a considerable influence on the culture during the centuries of Maya occupation. The important but gradual soil changes taking place were probably almost completely unnoticed by the Maya, owing to the slow and insidious nature of soil erosion and nutrient depletion. Sediments eroded from the upland soils at Tikal damaged the lowlands also. Large amounts of eroded sediments were deposited in drainageways and in Maya reservoirs. Sedimentation from soil erosion had a highly detrimental effect upon the limited Maya water supply in the urban setting.

At many locations in Maya areas, observations can be made like that shown in Figure 51, where erosion and sedimentation of the soils are associated with Maya decline and abandonment. Some of these soils observations have been documented by Harrison and Turner (1978). Archeologists digging into Maya mounds in Belize (Willey et al., 1965), for example, reported:

> In our excavations on the highest river terrace level...we noted that underlying the mounds at a level approximately 1.00 meter beneath the present ground level there was a stratum of black organically stained soil about 70 centimeters thick. This black stratum contrasted with the brown clays above and beneath it and was similar in appearance to the present humus stratum which forms the uppermost soil zone over the site. We noted that mounds whose construction began in Preclassic Period times had their bases directly on or slightly into the black soil stratum, while mounds whose construction began at a later

date had sterile brown clay intervening between the mound base and the black soil. It is tempting to ascribe this phenomenon to the following hypothetical explanation.

When the first Maya settlers arrived along the Belize Valley in Preclassic times, the uppermost terrace was heavily forested. As today, it was rarely, if ever, reached by flood waters, so that a substantial zone of humus, represented by the black soil stratum, could accumulate. When the land was cleared for habitation this humus accumulation was halted. At about the same time as the Belize Valley was being settled, Maya groups were also occupying the limestone country throughout the drainage basin of the river. As settlement of the region became more extensive and population increased, sufficiently large tracts of high forest were felled for agriculture and settlement to effect the moisture-holding and absorbing qualities of the forest cover as a whole. This meant that the rainy season runoff was increased in both quantity and intensity, resulting in more severe flooding downstream along the river. The uppermost river terrace at Barton Ramie now was occasionally flooded, accounting for the sterile alluvial deposit between the old humus zone and the bases of the later mounds. Increased flooding might also account for an apparently increasing tendency for higher house substructures during the later phases of the occupation of the site. After the region was abandoned or nearly abandoned, at the close of the New Town Phase, the forest as well as the river reverted to the conditions which had prevailed prior to the Maya occupation.

SARDIS

Sardis in western Turkey is another large archeological site that has had extensive soils investigations (Olson and Hanfmann, 1971). Sardis is a ruined city about 75 km east of Izmir in western Turkey, probably first settled in the third millennium B.C. The city has experienced droughts, earthquakes, famines, fires, floods, invasions, landslides, and sieges. It was the capital of the ancient kingdom of Lydia, the western terminus of the Persian royal road described by Herodotus, a center for administration under the Roman empire, and the metropolis of the province of Lydia in later Byzantine times. A strategic military location, its position on a main highway between the Anatolian plateau and the Aegean coast, and its access to the wide fertile plain of the Gediz river valley, all contributed to Sardis' importance.

Sardis was referred to in the *Iliad* (as Hyde), mentioned by the Greek poet Alcman about 650 B.C., and addressed as one of the Seven Cities of Asia in the Book of Revelation in the Bible. It was captured by the Cimmerians in the seventh century B.C., by the Persians in the sixth century B.C., by the Athenians in the fifth century B.C., and by Antiochus the Great in the third century B.C. It was destroyed by an earthquake in A.D. 17, but was rebuilt under Tiberius. The fort on the acropolis was handed over to the Turks in 1306. It was captured by Timur in 1402. The latest conflicts at Sardis were battles of the Greco-Turkish war in the first quarter of the twentieth century. Ruins at Sardis have been excavated since 1958 by a Cornell-Harvard expedition.

Soils of Sardis were of special interest in efforts to reconstruct the evolution of the urban community in relation to its environment. Soils were described and sampled according to the procedures outlined in this book (Fig. 64). In an amazing fashion, much of the history of Sardis is reflected in the soils. Most of the soils are Entisols and Inceptisols on steep slopes or in alluvial valleys, affected by landslides, erosion, and alluvial

FIGURE 64 *Describing the soil profile from a pit at Sardis. This soil formed in landslide materials, and has platy structure inclined downslope. Soil descriptions help in reconstruction of past events and ecological and environmental processes, and assist in predicting magnitude and direction of landslides of the future.*

FIGURE 65 *Part of the ruins at Sardis, with much of the visible building destroyed by soil erosion and landslides. The valley of the Gediz River can be seen in the distance to the north. The mountains were originally covered with forest, but most of the trees were cut down centuries ago.*

deposition. Soils on upper slopes have been overgrazed by goats and sheep for centuries; many of the ruins have had their foundations destroyed through the actions of erosion and landslides (Fig. 65). Forests originally protected the soils on the mountains, but the trees were cut for building constructions and to provide firewood for the Roman baths. In a sense, the civilizations destroyed themselves because of the ecological and environmental degradation; see Figure 66. In one place at Sardis, archeologists dug a trench 12 m deep through a Roman dump containing thousands of animal bones. Because of the overgrazing and subsequent soil erosion, Roman engineers were forced to design settling pots at close intervals in certain water lines at Sardis, so that soil sediments could be periodically removed. The Lydian occupation seems to have been most affected by flooding, and the most conspicuous landslides followed in the Greek and Roman periods. A stratum at one point had artifacts from 1300 B.C. buried more than 12 m below the modern soil surface. Although the floods and landslides were disastrous for the inhabitants of Sardis at the time, the deposits of soil materials were a boon to archeologists because they preserved some of the ruins from further destruction by nature and by human beings.

Many of the soils at Sardis have large contents of mica flakes, which contribute to sliding, especially when they are wet. The soil structure in these places (Fig. 64) not only records past soil movements, but also indicates possible directions and magnitudes of future landslides likely to follow major earthquakes. The splendor of ancient Sardis still evident (Fig. 66) gives cause for us to ponder our own future. Obviously, when the Romans built their huge bath complexes, they did not consider the ecology or sustained yield factors for the forests cut to provide heat for their water baths. Nor did the Greeks

FIGURE 66 *Roman bath building below the acropolis at Sardis. About 6 m of sediments have been deposited above the original base of the building. The forests protecting the mountains were cut to provide firewood to heat the water for the baths, and the resultant landslides and floods caused the alluviations that buried the structure.*

comprehend the effects of their goat herds on soil erosion. For their part, the Lydians overlooked the hazards of building on an alluvial floodplain.

What benefits can be obtained from evaluation of past civilizations on the soils of Sardis? Experiences at Sardis should emphasize to future generations that constructions on floodplains are likely to be damaged by floodwaters and alluvial deposits. Landslides, common to Sardis and many other parts of the world, can be predicted by internal and external soil characteristics. Soil erosion, whether at ancient Sardis or in modern urban construction areas, cannot extend over large areas for long periods without causing severe damage to structures and watersheds. The silting of water systems at Sardis was remarkably similar to some of our siltation problems in municipal water reservoirs. Erosion control is particularly critical for maintenance of soil fertility for food production as well as in environmental quality considerations. Guiding human activities toward consideration of ecology and natural soil landscape variations would be a positive step toward assuring that future civilizations will enjoy greater success and persist longer than did past civilizations. Interdisciplinary efforts to better understand the environment and its relationships to the welfare of inhabitants would contribute toward a more optimistic view of the future.

MEXICO CITY

Many examples through the ages illustrate effects of soil conditions upon human occupation of the land. The Aztecs built a large city before the Spanish conquest in the

FIGURE 67 *Building sinking into the unstable lacustrine soils at the Shrine of Guadalupe in Mexico City. This building was constructed about 1700 and had an inadequate foundation platform. A new basilica nearby has been constructed with more modern specifications for foundation support in the wet, spongy soils.*

vicinity of present Mexico City. It was built in a lacustrine basin and had elaborate systems of canals. Apparently, the Aztecs learned a great deal about soil engineering and drainage within the city, and also developed irrigation systems for farming around the city to sustain the large populations. Some of the problems of heavy constructions in the area are illustrated in Figure 67. Soil problems can be overcome with expensive engineering designs and other considerations, but it is often more economical and efficient to build and cultivate in harmony with the natural pattern of the soils. Cities and farms should be located in areas where soils are suitable as much as is possible, and the soil modifications necessary should also be in harmony with the environment as much as is feasible to be most effective over the long run.

NEGEV DESERT

In the Negev desert (Evenari et al., 1971), experiment stations have been established and interdisciplinary teams have conducted investigations into ancient relationships between human beings and the environment. The interdisciplinary teams included geologists, archeologists, agronomists, engineers, photogrammetrists, soil scientists, and historians. A farm at Avdat was reconstructed near the ancient city of Shivta to demonstrate that the desert environment could be made habitable once again with careful husbandry of the sparse resources. The Negev has been inhabited for many thousands of years. Paleolithic and Mesolithic peoples (before 11,000 B.C.) were mainly hunters. In the Neolithic (5000-4000 B.C.) the area was well populated, and people lived and farmed in the wadis and loess plains. For about 2,000 years much of the southern and central Negev was abandoned. Colonization was intense in the Bronze Age (about 2000 B.C.). By 1000 B.C. the Negev was part of the Israelite kingdom. Around the sixth century B.C. the Kingdom of Judah was destroyed by the Babylonians, and Bedouin nomads grazed the Negev with their flocks. From about 300 B.C. to A.D. 600 the Nabatean-Roman-Byzantine cultures occupied the desert, and the area prospered intermittently. Trade flourished in the Byzantine period, and caravan routes were established. The Arab conquests forced the area once again into decline, and farms and cities were abandoned. Soils and substrata in the Negev today contain artifacts of all these civilizations, and record much of the history of the military conquests and environmental changes at each site.

The soils of the Negev (Evenari et al., 1971) contained evidence that two systems of ancient agriculture (narrow terraced wadis and farm units with small watersheds) had a most rational and wise use of the available natural resources. The small plots fit into the natural soils landscapes without damaging the environment. When the agricultural systems were expanded into large catchments, however, the expectations were greater than the resources and resulted in "miscalculation leading to erosion, silting up, and destruction because this approach to cultivation of the desert was overambitious." The agricultural systems are another excellent illustration of the "tragedy of the commons"; when the yield expectations were modest, the area flourished, but when the yield demands were greater than the soil and water resources of the landscape could bear, the farming system collapsed.

The overgrazing that followed the destruction of the farming systems in the Negev also provides another example of the "tragedy of the commons." As Evenari et al. (1971) state: "When the Bedouin took over the Negev they mismanaged the desert in another way. They overexploited one of its main natural resources, the indigenous vegetation, by continuous overgrazing. They never made the slightest effort to restore some of the natural pastures which consist of the plant communities of the various desert ecosystems. Consequently plant and animal life deteriorated seriously."

RAJASTHAN DESERT

In the Rajasthan Desert of northwestern India, the Harappan civilization had a thriving culture about 2000-1700 B.C. Apparently, the climate was more humid then, but some believe that the aridity of today has been caused by past and present excessive human and animal populations. Meteorologists, archeologists, and other scientists have investigated the area for many years (Bryson and Murry, 1977). It appears that over-grazing and abuse of the vegetation have caused excessive soil erosion, especially from the strong winds. Dust in the air causes the moist atmosphere to cool and sink, instead of warming and rising and forming precipitation. The overgrazing has persisted to the present time, as Figure 68 illustrates. Cows are considered to be holy creatures in India, so that they are allowed to eat all the vegetation that they can reach. Human populations have destroyed the forests that once existed for firewood. Shortage of fuel dictates that cow dung must be now used as fuel for cooking, and consequently the fertility of the soils is further depleted. The vicious cycle of soil abuse is another example of the "tragedy of the commons" and will have increasing implications in the future in the social consequences.

FIGURE 68 *Scene in the Rajasthan Desert in northwestern India. This area once had prosperous fields, but the excessive human and animal populations have destroyed most of the vegetation. Dust in the atmosphere reduces the rainfall, so that the area has a persistent human-made desert. Revegetation of the soils could reverse the desertification process and make the desert productive once again.*

The archeological considerations for the most part give a pessimistic record of the abuse of soils and the social ramifications, but we must take these data seriously because of their sole witness over millennia of land abuse. For many, the archeological record is beyond comprehension because lifetimes and human memories are confined within a century or so. Future generations, however, will use the same soils through millennia that have already been abused for centuries or millennia in some cases. Fortunately, documentation of land abuse is becoming more widely available (Coates, 1972; Lowdermilk, 1975; Thomas, 1965). Increasing awareness of the importance of soils demands that soils be considered in strategic planning (Clawson et al., 1971). As Brown (1977) so eloquently states, soil and environmental resources and their management should be the most important considerations for the security of nations. Ultimately, contemporary constructions (Fig. 69), too, will become a part of the archeological record to be judged by future scientists. Fortunately, increasing data are becoming available about soils and their distributions in landscapes which could help wiser decisions to be made in the future than have been made in the past.

FIGURE 69 *Earthen wall threatened by desertification from wind erosion in Iran. Although the wall protects the irrigated field within its confines, the system is threatened with desertification due to land abuse in the external overgrazed hinterlands. This photograph is an illustration of the "tragedy of the commons."*

CHAPTER 11

PLANNING
FOR THE FUTURE

The world is a glorious bounty. There is more food than can be eaten if we would limit our numbers to those who can be cherished, there are more beautiful girls than can be dreamed of, more children than we can love, more laughter than can be endured, more wisdom than can be absorbed. Canvas and pigments lie in wait, stone, wood and metal are ready for sculpture, random noise is latent for symphonies, sites are gravid for cities, institutions lie in the wings ready to solve our most intractable problems, parables of moving power remain unformulated and yet, the world is finally unknowable.—How can we reap this bounty?

This quotation from McHarg (1971) illustrates very well the concept that design and development should be in harmony with nature, and that abundance is available for all if we would use our resources wisely. A new awareness is arising that national security should be redefined in terms of improving uses of natural resources (Brown, 1977), instead of squandering them. Many books and references are available to show examples of how soil surveys and other resource inventories have been well utilized, and to suggest procedures and techniques for improving land uses in the future (Bartelli et al., 1966; Beatty et al., 1979; Haans and Westerveld, 1970; Jarvis and Mackney, 1979; Olson, 1973; Simonson, 1974; Swindale, 1976; Vink, 1975; Young, 1976). Yet, in spite of all the data and technology available, soils continue to erode because farmers are too poor to invest in soil conservation practices (Fig. 70). Villages, towns, and cities continue to be built in the wrong places, or else are damaged by environmental abuse (Fig. 71). With application of our knowledge, environments can be improved and restored (Fig. 72). Needs for improvements in land use have some urgency for implementation because of the expanding populations and resultant ecological degradation and social strife.

FAO WORLD SOIL MAP

One of the most useful recent advances in general knowledge of soils of the world has been the publication of the FAO-UNESCO Soil Map of the World (FAO-UNESCO, 1970-1980). For the first time the soils of the world have been mapped with a unified scale and legend. The map has been published in 10 volumes in the decade extending from 1970 to 1980, at a scale of 1:5,000,000. The 10 volumes are:

Volume I   Elements of the legend
Volume II   North America

FIGURE 70 *Oblique aerial photograph of dendritic soil erosion in central India. Farmers here are so poor that they cannot invest in soil conservation practices. Thus, their soils erode and the productivity of the land declines.*

FIGURE 71 *Eroded village in Iran. Overgrazing and desertification in the uplands can also cause flooding and damage in villages in lowlands. Often, overgrazing increases runoff so that groundwater supplies are depleted. The landscape of Iran is dotted with thousands of abandoned villages destroyed by the "tragedy of the commons."*

FIGURE 72 *Exclosure area for excluding livestock in Iran. Where numbers of grazing animals are managed or drastically reduced, the vegetation can be much improved in a relatively short period of time. With livestock exclusion, erosion and desertification can be reduced and groundwater supplies increased. And the "tragedy of the commons" can be reversed in trend.*

Volume III	Mexico and Central America
Volume IV	South America
Volume V	Europe
Volume VI	Africa
Volume VII	South Asia
Volume VIII	North and Central Asia
Volume IX	South East Asia
Volume X	Australasia

The world soil volumes and maps represent a tremendous effort extending over many years, particularly in correlation of the soil map legend and agreement for the nomenclature of the soil terms. In addition to data on the soils, information is also provided in the volumes on climate, vegetation, geomorphology, lithology, and land use.

With completion of the World Soil Map, the Food and Agriculture Organization of the United Nations has shifted emphasis toward interpretation of the soil map units for practical purposes. At the general scale of 1:5,000,000, the soil map enables a delineation of "agro-ecological zones" to be made to obtain a first approximation of the "production potential of the world's land resources" and to help provide the data base necessary for global and regional planning of future development (Higgins, 1979). The soil map interpretation project for this effort deals with rainfed production potential at two levels of inputs for crops, including pearl millet, sorghum, rice, maize, wheat, phaseolus bean, soybean, cassava, sweet potato, white potato, and cotton.

With computers, the FAO study (Higgins, 1979) assembles the soil inventory of the extent and composition of the map units of the World Soil Map, by country. The climate inventory is overlaid on the soil map to delineate the resultant agro-ecological zones. The computer is then used to calculate the extent of soil units (by slope class, texture class, and map unit phase) for major climatic divisions and length of growing season zones (30-day intervals). The computer matches the climate inventory with the crop requirements and calculates the biomass and crop yields by lengths of growing periods where the crop requirements are met.

The soil requirements of crops are matched by rating the soil limitations for the individual crops at the two levels of inputs for the soil units, slope classes, texture classes, and phases of the soil map. The computer rates the agro-climate constraints to crop production for each area, and also calculates the production increases at different levels of inputs. A suitability classification is made for each area based on the predicted crop yields and the benefit/cost ratios of investment inputs necessary. The computer application of the soil limitation ratings thus systematically considers the soil characteristics, the crop requirements, the inputs necessary to overcome the limitations, and the economic feasibility of improving the production in the soil map unit areas. Some soil areas, obviously, can be improved for cropping much easier than others, and some will give greater returns with lesser investments than others. In some cases poorer soil areas will provide greater economic returns than better soil areas for small investments, depending upon the farming systems and the economic price structure.

An International Soil Museum has been established at Wageningen in the Netherlands (Schlicting, 1979). Soils mapped on the Soil Map of the World will be collected at the Museum, fully analyzed, and put on display for educational and research purposes. Reference materials for the FAO world soil map will be collected at the Museum, as well as other relevant information on the major characteristics and properties of the mapping units of the Soil Map of the World. The work of the staff of the International Soil Museum will be coordinated with the work of the Food and Agriculture Organization of the United Nations and with local soils work in different countries.

SOIL TAXONOMY

Soils work throughout the world has been greatly facilitated through publication of Soil Taxonomy (Soil Survey Staff, 1975). Now for the first time the format of a key for a detailed soil taxonomy exists, and it is being extensively used to improve soil correlations in many countries. Considerable linkage exists between the FAO-UNESCO Soil Map of the World and Soil Taxonomy (Fig. 27); in fact, although the nomenclature of the two soil classification systems are somewhat different, the Soil Taxonomy detailed criteria are the bases for the units of many diagnostic soil horizons and other soil categories of the FAO-UNESCO system. An office of International Soil Correlator has been established in the U.S. Department of Agriculture (funded by the U.S. Agency for International Development) to refine the criteria of Soil Taxonomy relevent to other countries, especially in the developing tropical part of the world.

For many years (since World War II) the World Soil Geography Unit of the Soil Conservation Service of the U.S. Department of Agriculture has collected soil surveys and compiled soil maps for military purposes. Recently, much of that work has been declassified and is currently being used in programs of the U.S. Agency for International Development. As part of this effort to make the work of the World Soil Geography Unit available, Orvedal (1977) published a bibliography of soils of the tropics and the Food and Agriculture Organization of the United Nations has published a summary of the Unified soil classification system for engineering purposes (FAO, 1973).

The general soil map of the world (Fig. 27) prepared by the World Soil Geography Unit summarizes the soil resources that are available to people in planning for the future; the legend is stated in terms of the classification of Soil Taxonomy (Soil Survey Staff, 1975). Alfisols are soils with gray to brown surface horizons, medium to high base supply, and subsurface horizons of clay accumulation; they are usually moist but may be dry during the warm season. Aridisols are soils with pedogenic horizons, low in organic matter, and dry more than 6 months of the year in all horizons. Entisols are soils without pedogenic horizons. Histosols are organic soils. Inceptisols are incipient soils that are usually moist, with pedogenic horizons of alteration of parent materials but not of accumulation. Mollisols are soils with nearly black, organic-rich surface horizons and high base supply. Oxisols are highly weathered soils of tropical and subtropical regions that have mostly gentle slopes on surfaces of great age. Spodosols are soils with accumulations of amorphous materials in subsurface horizons. Ultisols are soils that are usually moist with a horizon of clay accumulation and a low base supply. Vertisols are soils with high content of swelling clays and wide deep cracks in the dry season. All soils on the surface of the earth can be described in these general terms, and each can be placed in one of the 10 categories for broad global or regional planning purposes.

CRIES

The fullest use of soil information in the future will be dependent upon how well the data are integrated into the entire process of decision making. The Comprehensive Resource Inventory Evaluation System (CRIES) offers one technique by which the incorporation of soil information can take place (Putman, 1979). In the Dominican Republic, for example, the soil map information from the World Soil Geography Unit was integrated into a computer grid cell system (1 km^2) for regional planning. The soil map unit symbols have three parts: (1) the symbol of the dominant soils according to Soil Taxonomy, (2) the numerator of a fraction that shows the geomorphic landform, and (3) the denominator of a fraction that indicates the soil parent material or the material underlying the soil. The landforms of the Dominican Republic are dissected terraces (DT), fans at bases of mountains (F), floodplains (FP), karst topography (K), level plains (L), mountains (M), rolling and hilly terrain (RH), steep hills (S), terraces (T), undulating plains (U), undulating and rolling plains (UR), and swamps (WS). The underlying materials are unconsolidated alluvium (A), unconsolidated lacustrine alluvium (LA), limestone (LS), limestone and shale (LSS), unconsolidated marine sediments (M), mixed acid and basic igneous and metamorphic rocks (RB), and volcanic tuff (T). Double symbols indicate associations of groups of soils in the legend of the soil map. Thus, DOAa F/A–DOAa RH/LS has mostly variable kinds of Typic Camborthids on fans and uplands underlain by alluvium and limestone, and EOHj L/LA has mostly Ustic Torriorthents on level plains over lacustrine and alluvial materials.

Plant life zones were also used in the CRIES system, overlaid by computer onto the soil map. Thus, biotemperature, precipitation, and humidity were integrated with the soil map units to produce "resource production units"; these plant and soil boundaries divided the Dominican Republic into about 50 planning regions. "Production potential areas" were aggregated by computer to delineate areas "sufficiently homogeneous with respect to plant adaptability, potential productivity, and management requirements to be reliably depicted by unique agronomic and economic estimates for national and regional analysis and planning" (Putman, 1979). Many different kinds of economic data are input into the computer system to correlate with the "resource production units" and "production potential areas."

BENCHMARK PROJECT

Another project of considerable significance for the future to users of soil surveys is the Benchmark Soils Project (Silva and Beinroth, 1979) of the University of Puerto Rico and the University of Hawaii and other universities, agencies, and institutions. This study is also partly financed by the U.S. Agency for International Development and consists mainly of locating new agricultural experiment stations primarily to test soil-crop responses. With similar crops and management systems under similar climatic conditions the same soils (as classified in Soil Taxonomy) should produce the same results, within the defined limits and ranges. The "technology transfer" concept of this approach to soil and agronomic research would enable data to be extended from areas where experiments exist to other similar areas where data do not exist. Several stations of this network are located in Africa, Asia, and Latin America on Hydric Dystrandepts, Tropeptic Eutrustox, and Typic Paleudults. "Technology transfer" of these data can even be achieved across continents, where climate, management, and similar soil components can be carefully isolated and quantified.

SOIL QUANTIFICATION

In the future, soils will be quantified in more and more detail and soil maps will be made at larger scales with increasing intensity of field examinations. More farms, cities, project areas, and other locations will be planned and managed on the bases of soil maps and other natural resource inventory information. Computers will increasingly play a role in management and correlation of the resource inventories. Soil variability will be increasingly characterized, both within and between soil map units. Figure 73, for example, illustrates variable corn growth reflecting the different soil conditions; such variability is common in many places, especially where some soil factors are limiting (Beckett and Webster, 1971; Cipra et al., 1972).

The importance of understanding soil variability for management of areas of land in the future is illustrated in Tables 47 and 48. Table 47 is the result of statistical analyses of data like that presented in Table 14, but the computer analyses illustrate the number of samples likely to be necessary to characterize the fertility levels of the Conesus, Erie, Honeoye, Howard, Langford, and Volusia detailed soil map units (Table 47). From the laboratory and statistical analyses, the pH test is shown to be the most reliable. Analysis of only one sample for pH is likely to be characteristic of that soil map unit within ±20 percent of the pH number. Analyses of five samples for pH are likely to be characteristic of that soil map unit within ±10 percent. Organic matter characterization would require about three to seven samples for ±20 percent accuracy, and from 18 to 29 samples for ±10 percent accuracy. The test for ammonia nitrogen (NH_3-N) is most variable, and would require 1,978 samples for an accuracy level of ±10 percent on Howard soil map units—but thousands of soil samples are analyzed each year in New York State, so that

FIGURE 73   *Variable growth in cornfield in northern New York, caused by variable soil conditions. Differences in soil texture, structure, consistence, depth to fluctuating water table, erosion, and depth to bedrock make crop responses very different within these fields; the differences are especially apparent where the soil fertility levels are low. Drainage and improved liming and fertilizer management programs would be particularly helpful in raising the productivity of these areas.*

the collection and statistical analysis of these huge amounts of data require only the computer storage of the results of the laboratory analyses for a few years.

Table 48 is a similar statement on soil variability like that presented in Table 47, but the analyses are from soils in an alluvial floodplain in the Philippines. Techniques of chemical analyses are somewhat different in the two sets of data, but the variability is somewhat similar. The pH test in the Bicol River alluvium (Table 48) requires only one sample to characterize the reaction levels within ±20 percent, and only four to six samples to a ±10 percent accuracy. Organic carbon requires more samples for characterization of soil map units (7-82) in the Bicol River alluvium, in comparison with organic matter in the soils in glacial materials in New York State (3-29 samples required). These numbers have proved to be extremely valuable, because they emphasize the need for large amounts of data, and illustrate also the need for computer storage and manipulation of the results of the laboratory analyses. These data also show the relative reliability of a single laboratory determination and emphasize the need for careful collection of soil samples from soil map units described and delineated in the soil survey.

With the data provided by soil surveys and soil tests, planning for the future can be effectively done to improve soil performance. Figure 74 illustrates a soils landscape on a farm where the soils are highly contrasting, but the soil map is used as the base to plan a different use for each soil that should be managed differently. Soils in glacial outwash and alluvium in the valley can be intensively farmed and fertilized; they are very responsive

TABLE 47 *Number of samples necessary to estimate the mean of the population with different levels of accuracy using a 95 percent confidence interval (adapted from analyses of soil fertility samples from central New York State; Rogoff, 1976)*

Soil	Mean limits of accuracy (%)	Organic matter	pH	Exch H	P	K	Mg	Ca	Mn	Fe	Al	NO_3–N	NH_3–N	Zn	Soluble salts
Conesus	±10	19	5	33	290	117	116	171	49	424	547	471	219	114	172
	±20	6	1	10	72	29	29	43	12	106	137	118	55	28	43
Erie	±10	29	2	41	1018	111	42	102	40	172	675	637	49	68	320
	±20	7	1	10	255	28	10	25	10	43	169	159	12	17	80
Honeoye	±10	18	3	63	745	220	80	122	45	363	483	254	170	30	296
	±20	4	1	16	186	55	20	30	11	91	121	64	42	7	74
Howard	±10	26	4	26	400	186	176	67	184	252	162	353	1978	212	218
	±20	7	1	7	100	44	44	17	46	63	41	88	494	53	55
Langford	±10	5	3	42	1389	124	71	38	140	408	502	400	714	50	380
	±20	3	1	21	370	31	18	4	35	101	126	149	179	13	95
Volusia	±10	25	3	27	517	270	202	129	110	238	212	877	1834	81	900
	±20	6	1	7	129	68	50	32	28	59	53	219	458	20	225

TABLE 48 *Number of samples necessary from several soils to estimate the mean of the population with different levels of accuracy using a 95 percent confidence interval (adapted from analyses of soil samples in Bicol River alluvium from Luzon, Philippines; Concepcion, 1977)*

Soil property	Bantog ±10	Bantog ±20	Bigaa ±10	Bigaa ±20	Libmanan ±10	Libmanan ±20	Pamplona ±10	Pamplona ±20
pH	4	1	6	1	6	1	4	1
Organic carbon	82	20	29	7	79	20	61	19
P	327	81	217	54	184	45	249	63
K	44	11	72	18	54	13	50	13
Free Fe_2O_3	234	58	85	21	485	97	87	23
Exchangeable cations								
Ca + Mg	67	17	16	4	20	5	28	7
Na	148	42	87	21	116	29	208	49
K	357	86	2062	515	175	58	290	34
H	54	15	69	17	542	131	35	7
Cation-exchange capacity	32	8	12	3	7	2	14	3
Base saturation	17	4	2	1	7	2	6	1
Conductivity	128	32	40	10	182	45	77	19
Sand	196	49	81	20	230	57	120	30
Silt	66	16	95	11	27	7	24	6
Clay	14	3	3	1	7	2	10	2
Field capacity	12	3	5	1	7	2	8	5
Wilting coefficient	30	7	5	1	30	7	22	6
Available moisture	17	4	37	9	47	12	27	7
Settling volume	12	3	3	1	18	4	6	2

to good management. The upland soils in glacial till have problems with slope, dense subsoils, and shallowness to bedrock. Trees are the best crop for some of the soils, where erosion is a major problem. Contour cultivation and strip-cropping is very effective for erosion control on some of the gentler slopes, and some areas need drainage where seepage is a problem. Contrast the field in Figure 74 with the field in Figure 73; the variability of the corn plants in Figure 73 is mainly due to soil conditions, but the field would appear much more uniform when a higher fertility level and better soil drainage is achieved. Thus, the prosperity of a farm or region depends upon the soil characteristics, but management is also a critical factor in improvement of the soils for greater and more efficient production.

Improving uses of soils in the future will be much dependent upon more precise characterizations of moisture and temperature regimes in soils. More emphasis will be placed on seasonal fluctuations in soils affecting their use. Linkages between "real" and "inferred" soil properties will be brought more closely together. Soils are mapped in landscapes in accord with their soil profile characteristics, as Tables 49–51 illustrate. The Honeoye soil profile does not have mottles; consequently, it is classified as "well drained" in the soil drainage sequence (Fig. 31). Kendaia soil has mottles in the A2g horizon (Table 50) and is classified as somewhat poorly drained. Lyons soil is wet throughout the year, and has a black mucky A1 and distinct mottles in the B2g (with reduced olive colors in the C horizon) characteristic of poorly and very poorly drained soils. These soils occupy places in fields within several feet of each other; Honeoye soils occupy the higher knolls, Lyons soils are in the wet spots, and Kendaia soils are in the

FIGURE 74 *Soils landscape on a farm in New York State. Soils in the valleys have formed in glacial outwash and alluvium, and soils on the hills are formed in glacial till that is shallow to bedrock in places. Farm planning from soil maps helps the farmer to manage each soil for its best use, and control erosion on the steeper slopes. Effects of soil variability are reduced where the soils are improved through drainage and fertilization.*

TABLE 49 *Soil profile description of well-drained Honeoye soil (Glossoboric Hapludalf; fine loamy, mixed, mesic; adapted from Fritton and Olson, 1972)*

Ap—0 to 10 in., very dark brown (10YR 3/2) silt loam with few angular coarse fragments; moderate medium granular; friable; abundant fine fibrous roots; pH 6.2; clear wavy boundary

A2—10 to 15 in., brown (10YR 5/3) silt loam with pockets of dark-brown (10YR 3/2) earthworm casts and few coarse fragments of angular, subangular, and rounded shapes; weak granular in upper part grading into moderate medium to fine subangular blocky in lower part; very friable; common fine fibrous roots; pH 6.6; clear wavy boundary

B21—15 to 18 in., mosaic of dark-brown (10YR 4/3) silty clay loam and brown (10YR 5/3) and pale-brown (10YR 6/3) silty material with few angular coarse fragments; moderate medium subangular blocky; friable; common fine fibrous roots; pH 6.7; clear wavy boundary

B22—18 to 24 in., dark-brown (10YR 4/3) silty clay loam with common small pale silty pockets and strands; clayskins on some ped faces; moderate coarse subangular blocky; firm; few fibrous roots; pH 6.8; gradual irregular boundary

C—24 to 40+ inches, grayish-brown (2.5Y 4/4) gravelly loam with thin brown (10YR 3/3) clayskins and sprinkling of fine silt on upper side of some plates and few coarse fragments; weak moderate platy; firm; pH 7.9; strongly calcareous

Parent material:	Glacial till
Vegetation:	Grass in permanent pasture
Landform:	Ridge top
Slope:	About 1% convex

TABLE 50   *Soil profile description of somewhat poorly drained Kendaia soil (Aeric Haplaquept; fine loamy, mixed, mesic; adapted from Fritton and Olson, 1972)*

Ap—0–8 inches, very dark grayish-brown (10YR 3/2) fine-textured silt loam; moderate fine and medium crumb; very friable; many fine roots; pH 5.8; abrupt smooth boundary; 6 to 9 inches thick

A2g—8–17 in., pale-brown (10YR 6/3) silt loam with light grayish-brown (10YR 6/2) ped coatings and common medium distinct yellowish-brown (10YR 5/4) and grayish-brown (2.5Y 5/2) mottles; moderate fine and medium subangular blocky breaking into weak medium or thick platy; very thin discontinuous clayskins on peds; friable; many fine roots; pH 5.6; gradual wavy boundary; 5 to 9 inches thick

B2—17–24 in., dark grayish-brown (10YR 4/2) silty clay loam with very dark grayish-brown (10YR 3/2) ped faces and many fine distinct to prominent yellowish-brown (10YR 5/4–5/8) and light olive-gray (5Y 6/2) mottles; thick continuous clayskins that are pale brown (10YR 6/3) in upper part and very dark grayish-brown (10YR 3/2) in lower part; strong medium and coarse angular and subangular blocky; firm to very firm; pH 6.6; clear wavy boundary; 6–10 in. thick

Cg—24–33+ in., dark grayish-brown (2.5Y 4/2) and grayish-brown (2.5Y 5/2) silt loam; many fine faint light olive-brown (2.5Y 5/4) mottles; gray (10YR 6/1) ped coatings; moderate thick platy structure; firm; calcareous

Parent material:	Glacial till
Vegetation:	Grass in permanent pasture
Landform:	Gently undulating plain
Slope:	About 3% concave

intermediate landscape positions (with other soils also). The soils as observed, described, and mapped fit into the diagram shown in Figure 31 very well—and the water table is commonly predicted or "inferred" based on the soil profile observations and Munsell color notations.

Future measurements in soils will confirm and verify the observations and predictions of soil mapping to link the real and inferred soil properties. To test the hypothesis of soil mottling associated with wetness, for example, many holes were dug in soils and water-table fluctuations were measured weekly for a number of years (Fritton and Olson, 1972). Figures 75-77 summarize the results for the Honeoye, Kendaia, and Lyons soils. In Figures 75-77 the clear areas show the high and low measurements for each week, and specific data from several holes in each soil are plotted for a wet year and a dry year. Much variation is apparent, but over the long term the soil profile descriptions (Tables 49-51) are accurate representations of the actual water conditions in the soils. The zone of mottling (Fig. 31, Tables 49-51) locates the upper boundary of the fluctuating water table where the soil is saturated for a significant period each year. These data aid to establish the linkage between real and inferred soil properties, and characterize the

TABLE 51   *Soil profile description of poor and very poorly drained Lyons soil (Mollic Haplaquept; fine loamy, mixed, mesic; adapted from Fritton and Olson, 1972)*

A1—0–12 in., very dark gray (10YR 3/1) to black (10YR 2/1) mucky silt loam; weak fine crumb; very friable; pH 7.0; clear wavy boundary; 8–18 in. thick

B2g—12–23 in., gray (N 5/0) fine-textured silt loam; common medium distinct grayish-brown (2.5Y 5/2) and olive-brown (2.5Y 4/4) mottles; moderate medium and coarse subangular blocky; friable to slightly firm; pH 7.2; diffuse smooth boundary; 11–13 in. thick

C—23–36+ in., olive (5Y 4/4–5/4) silt loam with gray (N 5/0) vertical streaks 18–24 in. apart; massive; firm; calcareous

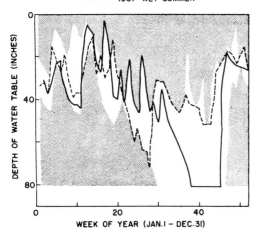

FIGURE 75 *Weekly measured depth to fluctuating water table in Honeoye soil over a period of several years (adapted from Fritton and Olson, 1972). The clear area (not shaded) locates the upper and lower limit for the water table for each week over the period of several years, and specific measurements are plotted for a wet year and a dry year.*

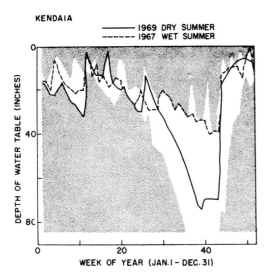

FIGURE 76 *Weekly measured depth to fluctuating water table in Kendaia soil over a period of several years (adapted from Fritton and Olson, 1972). The clear area (not shaded) locates the upper and lower limit for the water table for each week over the period of several years, and specific measurements are plotted for a wet year and a dry year.*

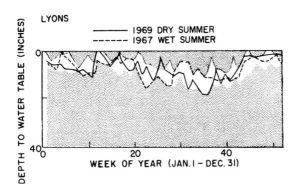

FIGURE 77 *Weekly measured depth to fluctuating water table in Lyons soil over a period of several years (adapted from Fritton and Olson, 1972). The clear area (not shaded) locates the upper and lower limit for the water table for each week over the period of several years, and specific measurements are plotted for a wet year and a dry year.*

seasonal variations that can be expected for each soil. Unfortunately, long-term observations of this nature are very expensive and laborious, so that careful planning and implementation in soil research is needed for these kinds of soil research projects. Detailed data such as these from research on a few soils can be readily extended to other areas of similar and related soils.

Soil surveys and projects for improving uses of soils in landscapes are in progress in nearly every country. Increasingly, in published reports, interpretations are receiving more emphasis to make soil surveys more useful. Computers are increasingly being used to help coordinate and correlate soil information to other data in planning of both rural and urban areas. Soil surveys are commonly tailor-made for farm planning and for project development (for city planning, watershed schemes, and irrigation and drainage areas). Soil scientists are serving as consultants and advisors to assist people in improving uses of soils and soil surveys. Increased training of soil scientists and others at colleges, universities, and regional centers is spreading knowledge widely about soils. In the future, ideas and concepts discussed in this book will play a major and increasing role in improving the fuller utilization of soils in landscapes.

CONCLUSIONS

Improving uses of soils in landscapes is one of the most important problems and challenges in management of natural resources. Soils are a vital part of every landscape, and they are described and mapped by soil surveys. Once soils are described and mapped, all kinds of interpretations can be made from the data about suitabilities, capabilities, and limitations of areas of land for different purposes. Soils data can be used for farm planning and city planning, for making land management to be most efficient and effective, and for making predictions about results of land use and abuse. Use of soils information is of vital concern to every person because of increasing pressures on the land. Increasing human populations demand that soils be used better in the future than they have been in the past. The crisis in world population is a real one (Hertzler, 1956), and abuse of soils in Haiti, Mexico, Iran, and elsewhere can ultimately have a serious effect upon other countries (Fig. 78). Increasingly, recognition is being given to the fact that soils are of strategic importance in all aspects of even human social and political behavior that deal with areas of land (Brown, 1977; Olson, 1980).

Fortunately, many uses have been made of soil surveys and many examples are available to guide people in planning for the future (Bartelli et al., 1966; Beatty et al., 1979; McHarg, 1971). Soil maps have been used for many years in farm planning in the United States—to help farmers control soil erosion and select the most efficient and productive cropping systems. Recently, the uses of soil surveys have expanded greatly, so that currently soil maps and data are used by agriculturalists, agronomists, assessors, botanists, conservationists, contractors, ecologists, economists, engineers, extension workers, foresters, geologists, groundwater experts, planners, politicians, public health officials, range managers, recreationists, teachers, wildlife specialists, and many others. Detailed soil maps are most useful because they are valuable for planning and developing specific tracts of land, but general soil maps are valuable also for regional and national planning and for teaching about soils. Computers and statistical methods are increasingly being used to make soil information more readily available, and to characterize variability and reliability of descriptions of soil map units.

FIGURE 78 *Overcrowding in a park in Mexico City. Too many people on small areas of land cause trampling of the vegetation, compaction of the soils, and lowering of the aesthetics and healthfulness of the environment. Better planning in accord with soil conditions could enable expansion of the park system in a more rural setting, providing of more and better facilities (picnic tables, waste disposal, etc.), and the general enhancement of the environment for the benefit of the human populations. Environmental degradation through excessive use has been termed the "tragedy of the commons" (Hardin, 1968). This park is a "commons" area, and no procedures have been imposed to restrict its use.*

Evaluation of prime farmland for preservation for agriculture into the future is one example of the application of soil maps on local, state, and national levels. Yield estimates of soil map units and land capability classes are commonly used as a base for taxation of farmlands. The uses for soil maps are almost unlimited, and are restricted only by the limits of human imagination and by accessory data needed to supplement the soil map units delineated in landscapes.

Use of soil information is enhanced with supplementary and interdisciplinary data. Thus, soil fertility tests, engineering soil tests, soil classification lab analyses, economic data, and all kinds of yield and performance information and correlations to soil map (landscape) units improve the application of soil surveys. Agricultural prosperity can be predicted when soil maps are correlated with economic data about farming (Olson and Hardy, 1967); subsurface sewage disposal in soils can be improved when seepage field designs are correlated to soils and recommendations are based on the soil characteristics (Huddleston and Olson, 1967). Even large urban developments can be sited and developed with assistance provided by soil information made in special investigations at large scale with deep borings (Olson and Marshall, 1968).

Much of the soil deterioration taking place is of a long-term insidious nature. Soil erosion is almost completely unnoticed even by farmers in most places, but it can lower productivity and have serious consequences within just a few years (Eckholm, 1976). When people exploit the land and the soils without limits, then productivity is apt to be lowered forever. The "tragedy of the commons" often occurs where the long-term aspects are disregarded—where arid regions are overgrazed and eroded; where parklands are trampled and compacted; where forests are clearcut without replanting; where urban development destroys the prime farmlands; where continuous row crops result in excessive soil losses; where soils and groundwaters are polluted by excessive dumping at waste disposal sites; or where drainage patterns are altered unknowingly in the uplands so that flood damage is caused in the lowlands. Soil is a "commons" resource that must be used by future generations, and no one individual or group of individuals has the right to damage the soil for short-term profit.

Figures 79 and 80 illustrate the insidious nature of farmland destruction. From 1965 (Fig. 79) to 1980 (Fig. 80) the prime farmland was replaced by a lake during gravel pit mining operations. Much more prime farmland will be destroyed in the future. Soil maps locate the best soils and describe how they can be most effectively used for many purposes. Obviously, gravel is needed for construction purposes, and society must weigh the value of prime farmland against using the soils for other purposes. Some gravel operations can be made to extend deeper, so that less farmland is consumed. Resultant lakes can be developed for recreational purposes. In any event, fullest use of soil maps and other resource inventory information will help to improve planning for the future.

Uses of soil surveys are being expanded in nearly every country (Olson, 1977). In Brazil, for example, soils in the Cerrado region southwest of Brasilia are naturally infertile (Fig. 81). The aluminum toxicity is reflected in the crop growth (Fig. 82), as well as in the natural vegetation. With improvements in soil management, however, crop yields can be greatly increased (Fig. 83). Use of soil maps in planning can benefit urban as well as rural populations (Fig. 84). The potential for soil improvements is enormous (Robertson et al., 1978) when soil maps and other information are effectively used in planning for the future to improve uses of soils in landscapes.

In summary, the positive philosophy of soil use for environmental improvement has been excellently stated (Dubos, 1972):

In this book I present examples to show that the inner structure of a given system—man [people], society, or place—exerts a governing influence on its further development. In nature, inanimate objects also have an inner structure which imposes a pattern on the changes they undergo even when they are transformed into artifacts by man [human beings].

If more were known of nature and of man [society], many hidden aspects of the natural world could be brought to light and incorporated into human life. Such creative incorporation could improve environmental and human quality. The surface of the earth can be profoundly altered without desecrating it.

Ideally each person—especially each child—should find in his [or her] physical and social environments stages on which to act out his [or her] life, in his [or her] own way.

FIGURE 79 *Cornfield on prime soils south of Syracuse. This photograph was taken in 1965 and illustrates the productivity of the soils for producing food into the future. Notice the small gravel pit operation in the middle part of the photo.*

FIGURE 80 *Lake formed from gravel pit operations, where prime farmland existed in 1965 (see Fig. 79). This photograph was taken in 1980, and illustrates the blatant nature of soil destruction. Most of the passers-by on this busy highway from 1965 to 1980 did not even notice the prime farmland disappearance, and even now few ponder the full implications of the land-use shift from farmland to water.*

FIGURE 81 *Stunted and twisted vegetation in the Cerrado region southwest of Brasilia. Oxisols here have a prolonged dry season and aluminum toxicity to plant roots. Vegetation growth is much dependent upon the soil conditions.*

FIGURE 82 *Variability in plant growth of* Crotolaria *crop in the Cerrado region southwest of Brasilia. Low levels of soil fertility, a prolonged dry season, and aluminum toxicity make crop production marginal if the soils are not improved.*

FIGURE 83 *Experimental plots on Oxisols near Brasilia. These soils are like those shown in Figures 81 and 82, but they have been improved with fertilization and careful management. Irrigation would increase yields even further. These soils have considerable potentials for increased production under improved management to serve better the expanding population of Brazil.*

FIGURE 84 *Shopping center construction in Brasilia. Soil information and maps helped in the initial location of the city (Feuer, 1956), and are increasingly being used in city planning and in improvement of the suburban areas. Improved production of soils in the surrounding areas will also benefit the urban population.*

CONVERSION FACTORS
FOR U.S. AND METRIC UNITS

To convert column 1 into column 2, multiply by:	Column 1	Column 2	To convert column 2 into column 1, multiply by:
		Length	
0.621	kilometer, km	mile, mi	1.609
1.094	meter, m	yard, yd	0.914
0.394	centimeter, cm	inch, in.	2.54
		Area	
0.386	kilometer^2, km^2	mile^2, mi^2	2.590
247.1	kilometer^2, km^2	acre, acre	0.00405
2.471	hectare, ha	acre, acre	0.405
		Volume	
0.00973	meter^3, m^3	acre-inch	102.8
3.532	hectoliter, hl	cubic foot, ft^3	0.2832
2.838	hectoliter, hl	bushel, bu	0.352
0.0284	liter	bushel, bu	35.24
1.057	liter	quart (liquid), qt	0.946
		Mass	
1.102	ton (metric)	ton (U.S.)	0.9072
2.205	quintal, q	hundredweight, cwt (short)	0.454
2.205	kilogram, kg	pound, lb	0.454
0.035	gram, g	ounce (avdp), oz	28.35
		Yield or Rate	
0.446	ton (metric)/hectare	ton (U.S.)/acre	2.24
0.892	kg/ha	lb/acre	1.12
0.892	quintal/hectare	hundredweight/acre	1.12
		Temperature	
$9/5$ ($^\circ$C + 32)	Celsius	Fahrenheit	$5/9$ ($^\circ$F − 32)
	−17.8°C	0°F	
	0°C	32°F	
	100°C	212°F	

GLOSSARY

These definitions were prepared especially for use by laypersons unfamiliar with soil science. Further details and more technical descriptions are available in the references cited in this book and in glossaries published by the Soil Science Society of America, Soil Conservation Society of America, and Food and Agriculture Organization of the United Nations.

Acid Acid soils have a measured pH below 6.5. The descriptive terms range from slightly acid (pH 6.1-6.5) to extremely acid (pH $<$ 4.4). Hydrogen dominates the exchange complex of acid soils, at the expense of calcium, magnesium, and potassium. Acid soils are generally infertile, but can be improved for crops by liming.

Aggregate An aggregate of soil is a collection of individual particles into clumps or structural shapes. Various binding forces are involved, and the aggregates generally become more distinct as the process of soil formation advances.

Agronomy An agricultural discipline concerned with field crops and soils.

Alluvial Deposited by streams.

Anion A negatively charged ion, such as SO_4^{2-}.

Arid region A desert or area with low rainfall.

Atterberg limits The Atterberg limits are the liquid limit, plastic limit, and plasticity index of a soil. The *plastic limit* is the moisture content at which a soil sample changes from a semisolid to a plastic state. The *liquid limit* is the moisture content at which a soil passes from a plastic to a liquid state. The *plasticity index*, which measures the plastic range of a soil, is the numerical difference between the liquid limit and the plastic limit of a soil sample.

Bajo A unique lowland swamp with soils that shrink and swell, under tropical rain forest, in northern Guatemala.

Base Positive ion as Ca^{2+} in soils that acts as nutrient for plants and neutralizer of soil acidity. $CaCO_3$ added to acid soils raises the pH, and generally has a favorable effect upon crop productivity.

Bedrock Unbroken solid rock overlaid in most places by soil or unconsolidated geologic materials.

Benefit/cost ratio A measure of the relationship between economic return and investment cost of a project. Soils in an irrigation project area should have a favorable benefit/cost ratio if an irrigation and drainage project is to be feasible.

Bulk density The weight of dry soil per unit bulk volume. Plant roots cannot penetrate soil layers with high bulk density.

Capability unit Soil areas (groupings of soil map units) with similar potentials and continuing limitations for cropping.

Carbonate A general term for compounds in soils with CO_3^{2-} ions, as $CaCO_3$ (calcium carbonate).

Cation A positively charged atom or ion in soils (Ca^{2+}, K^+, Na^+).

Channery Soil mass with 20–50 percent coarse fragments by volume of thin flat shape from 2 mm to 3 in. in size.

Chroma Relative purity or strength of a spectral color.

Clay Mineral soil particles less than 0.002 mm in size.

Consistence Degree and kind of cohesion and adhesion in soils and the resistance to deformation and rupture. Soil consistence changes with moisture content and should be measured when the soils are dry, moist, and wet.

Contour cultivation Tillage on slopes following the contour of the land, in contrast to up-and-down slope cultivation. This tillage retards runoff and erosion and increases infiltration of rainfall into the soil.

Crop The cultivated produce of soil areas.

Ecology The branch of biology dealing with the relationships between organisms and their environment.

Ecosystem A system formed by the interaction of a community of organisms with their environment.

Effluent Liquid flowing out or from a source. Septic tank effluent is the liquid material contaminated with human waste that is treated by leaching through the soil.

Eluvial Material lost from the upper part (A horizon) of the soil profile due to the action of leaching waters.

Erosion The process by which the soil surface is worn away, such as actions by water and wind. Upper horizons of soils are generally the most fertile, so that once a few inches of A horizon are lost, the productivity of the soil declines rapidly. Soil erosion is insidious, and proceeds at such a slow rate that it is unnoticed by most people.

Erosion factor Component of the soil-loss equation $A = RKLSCP$. The soil loss per unit area for each soil (A) is determined by the rainfall and runoff factor (R) $\times$ the soil erodibility factor (K) $\times$ the slope length factor (L) $\times$ the slope steepness factor (S) $\times$ the cover and management factor (C) $\times$ the support practice factor (P).

Fallow Land unplowed or unseeded for one or more growing seasons.

Fertility The capacity of the soil to supply nutrients in proper amounts for plant growth when other factors are favorable.

Fertilizer Substance used as an additive to enrich soils to improve their productivity. Common fertilizer materials include nitrogen, phosphorus, and potassium constituents, lime, and trace elements (copper, boron, zinc).

Fescue Any grass of the genus *Festuca*, some species of which are cultivated for pastures or lawns.

Fragipan A natural soil subsurface horizon with high bulk density relative to the horizons above, seemingly cemented when dry, but when moist showing a characteristic brittleness. It is low in organic matter, mottled, and slowly permeable. It never forms in calcareous materials or in soils high in sand or clay.

Frost heaving Movement in soils caused by alternate freezing and thawing under moist or wet conditions. Diurnal temperature fluctuations in temperate climates in the spring season causes ice lenses to form successively, forcing foundations and roadbeds out of line by heaving pressures from the ice formation.

Glacial till Material deposited and compacted directly by ice as the glacier advanced and retreated. Glacial till is generally compact, dense, and contains a large range of materials from boulders to clay fragments all mixed together without sorting or stratification. Most fragments in glacial till are not as rounded or smoothed as when deposited by stream action.

Gravel Rounded coarse fragments in soils that exceed 2 mm in diameter. An individual piece is a pebble. The term "gravel" refers to a mass of pebbles.

Hazard As used in this book, refers to those soil characteristics causing danger, peril, risk, or difficulty. Landslides and floods would impose hazards on the population of an area, and soil erosion depletes nutrients and topsoil on a long-term basis.

Hue Spectral (rainbow) color related to the dominant wavelength of the light.

Humus The dark organic material in soils, produced by the decomposition of vegetation and animal matter and essential to the fertility of the earth.

Hydrologic soil group Four soil groups (A, B, C, and D) are mapped to predict the hydrologic runoff potential. Group A has high infiltration rates and low runoff potential. Group B has moderate infiltration and moderate runoff. Group C has slow infiltration and rapid runoff. Group D has very slow infiltration and very rapid runoff potential.

Illuvial Material gained in the lower part (B horizon) of the soil profile due to the action of leaching waters.

K **value** A soil erodibility factor (in the soil loss equation) which is the soil loss rate per erosion index unit for a specified soil as measured on a unit plot of standard size.

Lacustrine Strata deposited in the bottom of a lake.

Landscape A panoramic view of environmental scenery that includes segments with distinctive kinds of soils.

Legumes Plants of the family *Leguminosae* that fix nitrogen in the soil from the air, have high protein content, and have seeds usually dehiscent by both sutures, thus dividing into two parts.

Limitations Relative magnitude of soil problems. Soils with slight limitations are relatively free of problems, or the problems are easily overcome. Soils with moderate limitations need to have the problems recognized, but the problems can be overcome with good management and careful design. Soils with severe limitations have severe problems that are great enough to make use of that soil questionable for the intended purpose.

Liquid limit The moisture content at which a soil passes from a plastic to a liquid state.

Loam A soil having an apparent relatively even mixture of different grades of sand, silt, and clay.

Loess A loamy deposit formed by wind, generally yellowish and calcareous, common in the Mississippi valley and in Europe and Asia.

Montmorillonite Clay mineral characterized by the ability to expand during absorption of large quantities of water.

Mottle Spots of different colors in a uniform matrix of soil. In many environments, mottling indicates intermittent wetness.

Munsell Soil Color Chart Loose-leaf notebook of soil color chip pages by which soil colors can be matched and identified in a standardized fashion and expressed in a notation of hue, value, and chroma (as 5YR 4/3, reddish brown).

Paleosol An old buried soil.

Ped An individual soil aggregate.

Pedology The science that deals with the study of soils.

Permeability The ease with which gases, liquids, or plant roots penetrate or pass through a given volume of soil.

pH The negative logarithm of the hydrogen-ion activity of a soil.

Plastic Quality of a soil that enables it to change shape continuously under the influence of an applied stress and to retain the impressed shape on removal of the stress.

Plastic limit Moisture content at which a soil sample changes from a semisolid to a plastic state.

Porosity The volume percentage of the total bulk of a soil volume not occupied by soil particles.

Resource Environmental attribute that can be utilized for the benefit of human beings and other organisms in the environment.

Rockiness Amount of soil surface occupied by bedrock outcrops.

Salt Soil constituent, including compounds formed by the replacement of one or more hydrogen atoms of an acid with elements or groups, which are composed of anions and cations, and which usually ionize in solution. Salts are usually highly soluble, and tend to accumulate in soils in lowlands and seepage spots in arid regions where groundwater accumulates and evaporates.

Sand Soil mineral particles from 2.0 to 0.5 mm in diameter.

Sanitary landfill Garbage burial in soil materials where the refuse is compacted and covered with a layer of soil each day. This technique is much more sanitary than open dumping and burning.

Septic tank System for sewage disposal that includes holding tank (for settling of solids) and a tile distribution system for effluent treatment by seepage through the soil. Septic tanks are commonly used for household waste disposal in rural or suburban housing areas where community waste treatment facilities are not available.

Sequential testing Technique of testing and monitoring soil behavior in sequences across landscapes. Soils differ within short distances in a logical and predictable pattern across terrain, and sequential testing enables soil performance and character to be evaluated and quantified. Common sequences that can be tested in soils landscapes include drainage, texture, slope, pH, fertility, land use, crop growth, yields, and many other soil and land attributes.

Sesquioxides Oxides containing three atoms of oxygen and two of another element, such as aluminum oxide (Al_2O_3) and iron oxide (Fe_2O_3).

Shrink–swell Capability of a soil to contract and expand in volume with drying and wetting.

Silt Soil mineral particles 0.002–0.05 mm in diameter.

Soil A soil is a uniquely defined segment of the landscape designated by a name (e.g., Volusia) and delineated by map boundaries into soil map units usually drawn on aerial photographs. Each soil is unique in its combination of characteristics and behavior. Each soil occupies its respective landscape position because of the environmental actions of climate, organisms, and relief acting upon the geologic materials through time. Thus, each soil is unique and predictable, and soil maps have a very great utility in helping to evaluate and predict soil behavior when different soils are used for different purposes. More than 11,000 individual soils have already been identified, defined, and mapped in the United States, and many more have been named in other countries.

Soil family Family is the fifth category in the hierarchical system of comprehensive soil classification of Soil Taxonomy. In the classification "Typic Hapludult; fine-loamy, mixed, mesic," the Family designation is "fine-loamy, mixed, mesic."

Soil group Great Group is the third category in the hierarchical system of comprehensive soil classification of Soil Taxonomy. In the classification "Typic Hapludult; fine-loamy, mixed, mesic," the Great Group designation is "Hapludult." Subgroup is the fourth category in the hierarchical system of Soil Taxonomy. In the classification "Typic Hapludult; fine-loamy, mixed, mesic," the Subgroup designation is "Typic Hapludult." Many groupings of soils are made for practical purposes, such as to indicate areas shallow to bedrock or places where excessive frost heaving poses a problem for roadbeds and foundations.

Soil horizon A layer of soil material approximately parallel to the land surface and differing from adjacent genetically related layers in physical, chemical, and biological properties.

Soil map A base map (usually an aerial photograph) with soil boundaries delineated on it locating the different kinds of soils and slopes with a symbol in each soil map unit, accompanied by a legend and description of the soils.

Soil order Order is the first (highest) category in the hierarchical system of comprehensive soil classification of Soil Taxonomy. In the classification "Typic Hapludult; fine-loamy, mixed, mesic," the Order designation is "Ultisol."

Soil profile A vertical cut exposing the various parts of the soil.

Soil series Series is the sixth (lowest) category in the hierarchical system of comprehensive soil classification of Soil Taxonomy. Names such as Volusia, Mardin, and Canandaigua are series names, and identify an individual soil with all its unique characteristics and behavior.

Soil Taxonomy A basic comprehensive system of soil classification for making and interpreting soil surveys. It was published in 1975 in a massive soil key document of 754 pages (Soil Survey Staff, 1975).

Soil wire Soil material with high amount of clay can be rolled into a "wire" of $1/8 - 1/4$ in. diameter when moist. This is one of the tests used in describing a soil.

Stoniness Amount of coarse fragments (rounded, 10-24 in. in diameter) in or on the soil.

Strip cropping Cropping on the contour in strips each several tens of meters wide. With strip cropping, dense cover crops such as alfalfa alternate with row crops such as corn to reduce soil erosion on sloping land.

Structure The aggregation of primary soil particles into compound particles (clusters of primary particles) which are separated from adjoining aggregates by surfaces of weakness.

Terrace Contour or nearly level embankment at right angles to the slope which impounds water and enables slow drainage of excessive runoff, thus conserving the soil from erosion. The height and width of a terrace depend upon the characteristics of the soil and slope and the management to be imposed on the land.

Texture Size and distribution proportion of mineral particles in the soil.

Thin section Slide for microscopic viewing taken through a soil volume, made by impregnating a soil aggregate with plastic, which hardens, and then can be cut with a diamond saw and ground to "thin" section. These are used to identify minerals and other small features in soils.

Topsoil Surface horizons of soils. The term has common usage to designate soil material suitable for landscape design (e.g., for growing lawns) when it can be spread over soil materials that are less suitable.

Tragedy of the commons Concept described by Hardin (1968) that human beings are incapable of managing their resources when population and exploitation are uncontrolled. The "commons" is a community pasture, park, forest, soil area, or other resource that can be used by all. The "commons" system works well when the community population is small, but productivity and environmental quality decline when the population expands and exploits the commons excessively. The "tragedy" is that the ecological system reverts to a low level as a result of overexploitation. Most human activities and problems (e.g., nuclear arms race, pollution, slums, wars) can be related to the concept of the "tragedy of the commons."

Trefoil Leguminous plant of the genus *Trifolium*, usually having digitate leaves of three leaflets and reddish, purple, yellow, or white flower heads, comprising the common clovers.

Unaggregated Soil materials not clumped together in peds. Unaggregated or disaggregated materials are common to dry sandy soils.

Value Relative lightness of color. Value is a function (approximately the square root) of the total amount of light.

Yield Soil giving forth or producing by a natural process or in return for cultivation.

REFERENCES

Ahn, P. M., 1978, A note on the apparent effect of soil texture on soil erodibility as deduced from the Wischmeier monograph; p. 11-12 *in* Technical Newsletter (no. 1, December 1978): Rome, Land and Water Development Division, Food and Agriculture Organization of the United Nations, 20 p.

Ambler, J. R., 1977, The Anasazi: Flagstaff, Ariz., Museum of Northern Arizona, 52 p.

Asphalt Institute, 1969, Soils manual for design of asphalt pavement structures: Manual Series 10: College Park, Md., The Asphalt Institute, 267 p.

Bartelli, L. J., 1978, Technical classification system for soil survey interpretation, p. 247-289 *in* Brady, N. C., ed., Advances in agronomy: New York, Academic Press, v. 30, 293 p.

Bartelli, L. J., Klingebiel, A. A., Baird, J. V., and Heddleson, M. R., eds., 1966, Soil surveys and land use planning: Madison, Wis., American Society of Agronomy, 196 p.

Beatty, M. T., Petersen, G. W., and Swindale, L. D., eds., 1979, Planning the uses and management of land: Agronomy Monograph 21: Madison, Wis., American Society of Agronomy, 1,028 p.

Beckett, P. H. T., and Webster, R., 1971, Soil variability: a review: Soils and Fertilizers, v. 34, p. 1-15.

Bell, F. F., and Springer, D. K., 1958, Cooperative studies in soil productivity and management in Tennessee: Journal of Soil and Water Conservation, v. 13, p. 156-157.

Brady, N. C., 1974, The nature and properties of soils: New York, Macmillan, 639 p.

Brown, J. B., 1963, The role of geology in a unified conservation program, Flat Top Ranch, Bosque County, Texas: Baylor Geological Studies Bulletin 5: Waco, Tex., Baylor University, 29 p.

Brown, L. R., 1977, Redefining national security: Worldwatch Paper 14: Washington, D.C., Worldwatch Institute, 46 p.

Bryson, R. A., and Murray, T. J., 1977, Climates of hunger: mankind and the world's changing weather: Madison, Wis., University of Wisconsin Press, 171 p.

Buol, S. W., and Davey, C. B., 1969, The soil, the tree: Journal of Soil and Water Conservation, v. 24, p. 149-150.

Buol, S. W., Hole, F. D., and McCracken, R. J., 1973, Soil genesis and classification: Ames, Iowa, Iowa State University Press, 360 p.

Cipra, J. E., Bidwell, O. W., Whitney, D. A., and Feyerhern, A. M., 1972, Variations with distance in selected fertility measurements of pedons of western Kansas Ustoll: Soil Science Society of America Proceedings, v. 36, p. 111-115.

Clawson, M., Landsberg, H. H., and Alexander, L. T., 1971, The agricultural potential of the Middle East: New York, American Elsevier, 312 p., soil maps.

Coates, D. R. ed., 1972, Environmental geomorphology and landscape conservation: Volume I: Prior to 1900 (Benchmark Papers in Geology), Stroudsburg, Pa., Dowden, Hutchinson, & Ross, 485 p.

Concepcion, R. N., 1977, Soil variability in Bicol River alluvium: M.P.S. thesis: Cornell University, 129 p.

Decker, G. L., Nielsen, G. A., and Rogers, J. W., 1975, The Montana automated data processing system for soil inventories: Research Report 89: Bozeman, Mont., Department of Plant and Soil Science, Montana State University, 77 p.

Dubos, R., 1972, A god within: a positive philosophy for a more complete fulfillment of human potentials: New York, Scribner's, 325 p.

Eckholm, E. P., 1976, Losing ground: environmental stress and world food prospects: New York, W. W. Norton, 223 p.

Edwards, R. D., Rabey, D. F., and Kover, R. W., 1970, Soil survey of Ventura area, California: University of California Agricultural Experiment Station and Soil Conservation Service/U.S. Department of Agriculture: Washington, D.C., U.S. Government Printing Office, 151 p., 50 soil map sheets.

Evenari, M., Shanan, L., and Tadmor, N., 1971, The Negev: the challenge of a desert: Cambridge, Mass., Harvard University Press, 345 p.

FAO, 1977, Soil survey interpretation for engineering purposes: Soils Bulletin 19: Rome, Food and Agriculture Organization of the United Nations, 24 p.

FAO, 1974, Approaches to land classification: Soils Bulletin 22: Rome, Food and Agriculture Organization of the United Nations, 120 p.

FAO-UNESCO, 1970–1980, Soil map of the world: Prepared by the Food and Agriculture Organization of the United Nations, Rome: Paris, UNESCO, 10 v., 18 soil map sheets, scale: 1:5,000,000.

Feuer, R., 1956, An exploratory investigation of the soils and agricultural potential of the soils of the future Federal District in the Central Plateau of Brazil: Ph.D. thesis: Cornell University, 432 p.

Fly, C. L., and Romine, D. S., 1964, Distribution patterns of the Weld-Rago soil association in relation to research planning and interpretation. Soil Science Society of America Proceedings, v. 28, p. 125-130.

Folk, R. L., 1951, A comparison chart for visual percentage estimation: Journal of Sedimentary Petrology, v. 21, p. 32-33.

Foster, G. R., ed., 1977, Soil erosion: prediction and control: Special Publication 21: Ankeny, Iowa, Soil Conservation Society of America, 393 p.

Fritton, D. D., and Olson, G. W., 1972, Depth to the apparent water table in 17 New York soils from 1963 to 1970: New York's Food and Life Sciences Bulletin 13: Ithaca, N.Y., Agricultural Experiment Station, Cornell University, 40 p.

Greweling, T., and Peech, M., 1965, Chemical soil tests: Bulletin 960: Ithaca, N.Y., Cornell University Agricultural Experiment Station, New York State College of Agriculture, 60 p.

Grice, D. G., Green, W., and Richardson, W., 1965, Soil survey of Hockley County, Texas: Texas Agricultural Experiment Station and Soil Conservation Service/U.S. Department of Agriculture: Washington, D.C., U.S. Government Printing Office, 65 p., 70 soil map sheets.

Haans, J. C. F. M., and Westerveld, G. J. W., 1970, The application of soil survey in The Netherlands: Geoderma, v. 4, p. 279-309.

Hach, 1973, Instruction manual for Hach soil analysis laboratory Model SA-1: Ames, Iowa, Hach Chemical Company, 32 p.

Hardin, G., 1968, The tragedy of the commons: Science, v. 162, p. 1243-48.

Harrison, P. D., and Turner, B. L., II, 1978, Pre-Hispanic Maya agriculture: Albuquerque, N. Mex., University of New Mexico Press, 414 p.

Henao, J., 1976, Soil variables for regressing Iowa corn yields on soil, management, and climatic variables: Ph.D. thesis: Iowa State University, 315 p.

Hertzler, J. O., 1956, The crisis in world population: Lincoln, Nebr., University of Nebraska Press, 279 p.

Higgins, G. M., 1979, The "agro-ecological zones" project of FAO: Bulletin of the International Society of Soil Science, v. 55, p. 33-34.

Huddleston, J. H., and Olson, G. W., 1967, Soil survey interpretation for subsurface sewage disposal: Soil Science, v. 104, p. 401-409.

Jarvis, M. G., and Mackney, D., 1979, Soil survey applications: Technical Monograph 13: Harpenden, Herts, England, Soil Survey, Rothamsted Experimental Station, 197 p.

Klausner, S. D., and Reid, W. S., 1979, A summary of soil test results for field crops grown in New York during 1977-1978: Cornell Agronomy Mimeo 79-21: Ithaca, N.Y., Cornell University, 15 p.

Kling, G. F., and Olson, G. W., 1975, Role of computers in land use planning: Information Bulletin 88: Ithaca, N.Y., New York State College of Agriculture and Life Sciences, Cornell University, 12 p.

Klingebiel, A. A., and Montgomery, P. H., 1966, Land capability classification: USDA Agriculture Handbook 210: Washington, D.C., U.S. Government Printing Office, 21 p.

Lathwell, D. J., and Peech, M., 1973, Interpretation of chemical soil tests: Bulletin 995: Ithaca, N.Y., Cornell University Agricultural Experiment Station, New York State College of Agriculture and Life Sciences, 40 p.

Loehr, R. C., ed., 1977, Land as a waste management alternative: Proceedings of the 1976 Cornell Agricultural Waste Management Conference: Ann Arbor, Mich., Ann Arbor Science Publishers, 811 p.

Lopes, A. S., and Cox, F. R., 1977, Cerrado vegetation in Brazil: an edaphic gradient: Agronomy Journal, v. 69, p. 828-831.

Lowdermilk, W. C., 1975 (rev.), Conquest of the land through 7,000 years: Agriculture Information Bulletin 99: Soil Conservation Service/U.S. Department of Agriculture: Washington, D.C., U.S. Government Printing Office, 30 p.

Lytle, S. A., McMichael, C. W., Green, T. W., and Francis, E. L., 1960, Soil survey of Terrebonne Parish, Louisiana: Louisiana Agricultural Experiment Station and Soil Conservation Service/U.S. Department of Agriculture: Washington, D.C., U.S. Government Printing Office, 43 p., 140 soil map sheets.

Malo, D. D., and Worcester, B. K., 1975, Soil fertility and crop responses at selected landscape positions: Agronomy Journal, v. 67, p. 397-401.

Martini, J. A., 1977, A field method for soil test calibration in developing countries: Soil Science, v. 123, p. 165-170.

McBee, C. W., Fleming, E. L., Hamilton, V. L., and Sallee, K. H., 1961, Soil Survey of Hamilton County, Kansas: Kansas Agricultural Experiment Station and Soil Conservation Service/U.S. Department of Agriculture: Washington, D.C., U.S. Government Printing Office, 43 p., 84 soil map sheets.

McHarg, I. L., 1971, Design with nature: Garden City, N.Y., Doubleday, 198 p.

Montgomery, P. H., and Edminster, F. C., 1966, Use of soil surveys in planning for recreation, p. 104-112 in Bartelli, L. J., et al., Soil surveys and land use planning: Madison, Wis., American Society of Agronomy, 196 p.

Munn, L. C., Nielsen, G. A., and Mueggler, W. F., 1978, Relationships of soils to mountain and foothill range habitat types and production in western Montana: Soil Science Society of America Journal, v. 42, p. 135-139.

Munsell Color Company, 1954, Munsell Soil Color Charts: Baltimore, Md., 10 p. with color chips.

Neeley, J. A., Giddings, E. B., and Pearson, C. S., 1965, Soil survey of Tompkins County, New York: U.S. Department of Agriculture/Soil Conservation Service in cooperation with Cornell University Agricultural Experiment Station: Washington, D.C., U.S. Government Printing Office, 241 p., 38 soil map sheets.

Nelson, L. A. et al., 1963, Detailed land classification—island of Oahu: Land Study Bureau Bulletin 3: Honolulu, Hawaii, University of Hawaii, 141 p.

Nichols, J. D., and Bartelli, L. J., 1974, Computer-generated interpretive soil maps: Journal of Soil and Water Conservation, v. 29, p. 232-235.

Nobe, K. C., Hardy, E. E., and Conklin, H. E., 1960, The extent and intensity of farming in western New York state: Economic Land Classification Leaflet 7: Ithaca, N.Y., New York State College of Agriculture, Cornell University, 12 p.

Olson, G. W., 1973, Improving uses of soils in Latin America: Geoderma, v. 9, p. 257-267.

——, 1976, Criteria for making and interpreting a soil profile description: a compilation of the official U.S. Department of Agriculture procedure and nomenclature for describing soils: Kansas Geological Survey Bulletin 212: Lawrence, Kans., University of Kansas, 47 p.

——, 1977, The soil survey of Tikal: Cornell Agronomy Mimeo 77-13: Ithaca, N.Y., Department of Agronomy, Cornell University, 82 p.

——, 1977, Using soil resources for development in Latin America: Cornell International Agriculture Bulletin 31: Ithaca, N.Y., Cornell University, 32 p.

——, 1979, The "state-of-the-art" in linking soil fertility work to the soil survey: Soil Conservation Service, U.S. Department of Agriculture/Cornell University: Ithaca, N.Y., Agronomy Mimeo 79-29, Department of Agronomy, Cornell University, 143 p.

——, 1980, The Iranian situation according to the soils: Garden, v. 4, p. 20-24.

——, and Hanfmann, G. M. A., 1971, Some implications of soils for civilizations: New York's Food and Life Sciences, v. 4, p. 11-14.

——, and Hardy, E. E., 1967, Using existing information to evaluate and predict agriculture in a region: Journal of Soil and Water Conservation, v. 22, p. 62-66.

——, and Marshall, R. L., 1968, Using high-intensity soil surveys for big development projects: a Cornell experience: Soil Science, v. 105, p. 223-231.

——, and Warner, J. W., 1974, Engineering soil survey interpretations: Information Bulletin 77: Ithaca, N.Y., New York State College of Agriculture and Life Sciences, Cornell University, 8 p.

———, Witty, J. E., and Marshall, R. L., 1969, Soils and their use in the five-county area around Syracuse: Miscellaneous Bulletin 80: Ithaca, N.Y., New York State College of Agriculture, Cornell University, 100 p., map.

Orvedal, A. C., 1977, Bibliography of soils of the tropics: Volume II. Tropics in general and tropical South America: Technical Series Bulletin 17: Washington, D.C., Office of Agriculture, Technical Assistance Bureau, Agency for International Development, 245 p.

Orvedal, A. C., and Edwards, M. J., 1941, General principles of technical grouping of soils: Soil Science Society of America Proceedings, v. 6, p. 386-391.

PCA, 1973, PCA soil primer: Engineering Bulletin: Skokie, Ill., Portland Cement Association, 40 p.

Putman, J. W., 1979, The Comprehensive Resource Inventory and Evaluation System: a Caribbean experience, p. 155-164 *in* Soil resource inventories and development planning: proceedings of a workshop organized by the Soil Resource Inventory Study Group at Cornell University, December 11-15, 1978: Agronomy Mimeo 79-23: Ithaca, N.Y., Cornell University, 332 p.

Rivera, L. H., Frederick, W. D., Farvis, C., Jensen, E. H., Davis, L., Palmer, C. D., Jackson, L. F., and McKinzie, W. E., 1970, Soil survey of the Virgin Islands of the United States: Soil Conservation Service, U.S. Department of Agriculture: Washington, D.C., U.S. Government Printing Office, 80 p., 32 soil map sheets.

Robertson, L. S., Warncke, D. D., and Mokma, D. L., 1978, Test levels in profiles of two soils producing world-record corn yields: Research Report 363: East Lansing, Mich., Michigan State University, 4 p.

Robinette, C. E., 1975, Corn yield study on selected Maryland soil series: M.S. thesis: University of Maryland, 192 p.

Rogoff, M. J., 1976, Soil ratings for farming in central New York State: M.S. thesis: Cornell University, 195 p.

Sarkar, P. K., Bidwell, O. W., and Marcus, L. F., 1966, Selection of characteristics for numerical classification of soils: Soil Science Society of America Proceedings, v. 30, p. 269-272.

Schlicting, E., 1979, The International Soil Museum officially opened: Bulletin of the International Society of Soil Science, v. 55, p. 35-39.

SCS, 1969, Know your land: narrative guide, with photographs and captions, and color slide set: Soil Conservation Service, U.S. Department of Agriculture: Washington, D.C., U.S. Government Printing Office, 12 p.

SCS, 1973 (revised), What is a farm conservation plan?: PA-629, Soil Conservation Service, U.S. Department of Agriculture: Washington, D.C., U.S. Government Printing Office, 8 p.

Shaw, C. F., 1928, A definition of terms used in soil literature: First International Congress of Soil Science Proceedings and Papers, v. 5, p. 38-64.

Shiflet, T. N., 1972, Analysis of a data system for the collection, storage, and retrieval of rangeland resource information: Ph.D. thesis: University of Nebraska, 199 p.

Silva, J. A., and Beinroth, F. H., 1979, Development of the transfer model and Soil Taxonomy interpretations on a network of three soil families: Benchmark Soils Project, Progress Report 2, January 1978-June 1979: Department of Agronomy and Soil Science, College of Tropical Agriculture and Human Resources, University of Hawaii/Department of Agronomy and Soils, College of Agricultural Sciences, University of Puerto Rico, 81 p.

Simonson, R. W., ed., 1974, Non-agricultural applications of soil surveys: Amsterdam, Elsevier, 178 p.

Smith, H., 1976, Soil survey of the District of Columbia: Soil Conservation Service, U.S. Department of Agriculture: Washington, D.C., U.S. Government Printing Office, 114 p., 17 soil map sheets, 1 general soil map.

Soil Survey Staff, 1962 (reissued), Soil survey manual: USDA Agriculture Handbook No. 18: Washington, D.C., U.S. Government Printing Office, 503 p.

Soil Survey Staff, 1967, Soil survey laboratory methods and procedures for collecting soil samples: Soil Survey Investigations Report No. 1: Soil Conservation Service, U.S. Department of Agriculture, Washington, D.C., U.S. Government Printing Office, 50 p.

Soil Survey Staff, 1971, Guide for interpreting engineering uses of soils: Soil Conservation Service, U.S. Department of Agriculture: Washington, D.C., U.S. Government Printing Office, 87 p.

Soil Survey Staff, 1975, Soil Taxonomy: a basic system of soil classification for making and interpreting soil surveys: USDA Agriculture Handbook 436: Washington, D.C., U.S. Government Printing Office, 754 p.

Sopher, C. D., and McCracken, R. J., 1973, Relationships between soil properties, management practices, and corn yields on South Atlantic coastal plain soils: Agronomy Journal, v. 65, p. 595-599.

Swindale, L. D., ed., 1976, Soil resource data for agricultural development: Honolulu, Hawaii, Hawaii Agricultural Experiment Station, College of Tropical Agriculture, University of Hawaii, 306 p.

Thomas, W. L., Jr., ed., 1965 (reprinted), Man's role in changing the face of the earth: Chicago, University of Chicago Press, 1193 p.

Vink, A. P. A., 1975, Land use in advancing agriculture: Berlin, Springer-Verlag, 394 p.

Wallen, V. R., and Jackson, H. R., 1978, Alfalfa winter injury, survival, and vigor determined from aerial photographs: Agronomy Journal, v. 70, p. 922-924.

Westin, F. C., 1976, Geography of soil test results: Soil Science Society of America Journal, v. 40, p. 890-895.

Willey, G. R. et al., 1965, Prehistoric Maya settlements in the Belize Valley: Cambridge, Mass., Papers of the Peabody Museum of Archeology and Ethnology, Harvard University, v. 54, 589 p., map.

Wischmeier, W. H., 1966, Relation of field-plot runoff to management and physical factors: Soil Science Society of America Proceedings, v. 30, p. 272-277.

Wischmeier, W. H., and Smith, D. D., 1978, Predicting rainfall erosion losses—a guide to conservation planning: USDA Agriculture Handbook 537, Washington, D.C., U.S. Government Printing Office, 58 p.

Young, A., 1976, Tropical soils and soil survey: Cambridge, Cambridge University Press, 468 p.

INDEX